OPEN UP

The Neurospiritual Guide to Your Enneagram Type

Christophe Morin, Ph.D.

Author of OPEN, The Serenity Code, and The Persuasion Code

ABSTRACT

OPEN UP empowers you to transcend habitual patterns and cultivate profound self-awareness to exploit the self-healing properties of your brain. It is based on the *OPEN* neurospiritual model (Morin, 2025), which integrates neuroscience, neuroplasticity, and spiritual traditions. *OPEN UP* leverages the Enneagram's wisdom to suggest actionable strategies to eliminate the toxic effects of stress, anxiety, depression, addiction, and trauma (SADAT). The nine Enneagram types reflect ego-driven survival strategies that can significantly reduce consciousness and neuroplasticity. However, the neurospiritual model empowers each type to manifest higher virtues and achieve self-realization.

At the heart of this transformation is a sacred journey toward wholeness and transcendence. *OPEN UP* illuminates how you can move beyond your limitations by embracing your Holy Idea—the archetypal truth and virtue associated with your type. It emphasizes cultivating a temenos, or sacred space, where personal growth can unfold in safety and alignment. Whether physical or psychological, this temenos serves as a container for integrating shadow aspects, confronting fears, and unlocking a more expansive sense of self. In essence, *OPEN UP* helps you:

- Understand how your ego defenses limit personal growth and reprogram these through neuroplasticity.
- Foster greater self-acceptance and authenticity.
- Access a harmonious state where life unfolds purposefully and joyfully.

The book recommends tailored *OPEN rituals* for each type, enabling you to overcome habitual challenges and thrive. Ultimately, *OPEN UP* invites you to embark on a transformative path toward healing and fulfillment, supported by tools to cultivate resilience, consciousness, and flow.

Testing for your type is free for purchasers of the book and can be done at https://su.vc/nfyybfft

I dedicate this book to my wife, sons, mother, father, brother, sisters, and all the beautiful souls that have supported my splendid life journey.

Published by Depth Insights, Honolulu, Hawaii.

Limit of Liability/Disclaimer of Warranty: While the publisher and author have used their best efforts in preparing this book, they make no representations or warranties concerning the accuracy or completeness of the contents of this book and expressly disclaim any implied warranties of merchantability or fitness for a particular purpose. Sales representatives or written sales materials may create or extend no warranty.

The information shared in OPEN UP is for educational and informational purposes only and should be used solely as a self-help tool. By reading this book, you acknowledge that Dr. Morin does not provide therapy or counseling services. Dr. Morin serves solely as a coach and mentor, providing only educational and informational coaching. Education and coaching services are not a substitute for professional mental health care or medical care and, therefore, are not intended to diagnose, treat, or cure any mental health or medical conditions.

Some images featured in his book are licensed from Istockphoto.com and Salesbrain. Others are generated from AI.

ISBN: 978-0-9979550-6-4

Praises for *OPEN*

In this timely and useful book, Dr. Christophe Morin, a psychologist and neuroscientist, explains in simple terms how you can use neuroplasticity to overcome mental problems such as stress, anxiety, depression, addictions, and trauma, raise your consciousness, and experience peace, joy, and a sense of purpose. The innovative healing approach presented in OPEN is based on extensive scientific evidence and the author's own personal journey. I highly recommend this work.

Dr. Mario Beauregard, Ph.D., Neuroscientist, co-author of *The Spiritual Brain*, author of *Expanding Reality*.

Read this book with an open mind and curiosity. Suspend judgment and watch what happens as Dr. Morin takes you on a journey that will resonate and articulate how you can enhance your daily life. The insights, raw vulnerability, and research behind this book will enlighten any reader.

Dr. Russ Riendeau, Ph.D., Mayo Clinic Certified Wellness Coach

This is Dr. Morin's best work, unlike anything I have read. He is open about his trauma in a way that is both honest and respectful. The way he weaves his personal story with an intense psychedelic-cultural experience and then explains that science is pure genius. You will not be able to stop this intellectual feast.

Dr. Ian McCulloh. Ph.D., LTC (Retired) Associate Professor, Johns Hopkins University

This book is a fascinating journey into the inner healing potential of our brain and soul. The title says it all: Life is much better when we embrace our ability to reprogram toxic habits and live free of stress and anxiety.

Dr. Steve Feinberg, Ph.D. Neurostrategist and Executive Coach, author of *Do what others say can't be done* and *The Advantage Maker*

Dr. Morin offers a new way of being for us all. Sharing his personal journey and rigorous scientific insights in such a compelling manner is more than a creative act. He reminds us that humanity needs plasticity more than ever.

Joe Tafur MD, author of *The Fellowship of the River: A Medical Doctor's Exploration into Traditional Amazonian Plant Medicine*

Table of Contents

PREFACE

Writing OPEN was a profoundly transformative experience that illuminated the intricate interplay of neuroscience, spirituality, and personal growth. However, as the book unfolded, I grew convinced that there was more to be explored—a piece of the puzzle that demanded a voice. This is where the concept of OPEN UP: The Neurospiritual Guide to Your Enneagram Type came into focus. The Enneagram is a powerful way to bridge neuroplasticity insights with the intricate map of human ego, fear, and potential growth.

Understanding your Enneagram type offers insight into the habitual thought patterns, emotional responses, and behaviors that shape how you adapt to life's challenges. Each type has its specific defense mechanisms and challenges, which, if left unexamined, can lead to maladaptive coping strategies like avoidance or denial. For instance, a Type 9 (Peacemaker) may avoid conflict and disengage to maintain peace, which can result in a lack of assertiveness or emotional stagnation.

The Enneagram is an ideal framework to explore these patterns, illuminating how deeply ingrained personality tendencies shape our responses to adversity and our paths to healing. It is a credible diagnostic and developmental tool that links ego-driven survival strategies with higher spiritual virtues. Each type's unique challenges and strengths align beautifully with the OPEN model's call to raise consciousness, rewrite personal narratives, and reprogram life habits.

Meanwhile, writing OPEN UP allowed me to contextualize the importance of sacred spaces (temenos) where you can integrate shadow aspects and foster authenticity. These spaces are crucial for transcending the limiting narratives of the ego and embracing the Holy Idea associated with each Enneagram type. This process dovetails seamlessly with the neurospiritual principles of the OPEN framework. In essence, OPEN UP enriches OPEN by adding depth to understanding personality dynamics and their influence on consciousness and neuroplasticity. It brings to light the transformative potential inherent in each individual's journey through their type, offering practical, type-specific rituals and practices to cultivate resilience, joy, and flow. This synergy of OPEN's scientific rigor and the Enneagram's timeless wisdom creates a roadmap for holistic healing and profound personal growth.

I genuinely believe that with the help of OPEN UP, you can identify and actively rewire your type's unconscious coping mechanisms that do not serve you. Ultimately, OPEN UP is an invitation to transform challenges into opportunities, rewrite limiting narratives, and live a balanced, intentional, and authentically connected life. OPEN UP guides you to create a life rooted in growth, flow, and self-discovery.

Enjoy!

Dr. Christophe Morin

Kona, Hawaii March 1, 2025

PS: While I highly encourage you to read OPEN before reading OPEN UP, I've provided enough context about the book in the following few pages to ensure you can still benefit from its teachings even if you haven't read it yet or do not plan to do so.

Summary of Introduction: The Power of Being OPEN

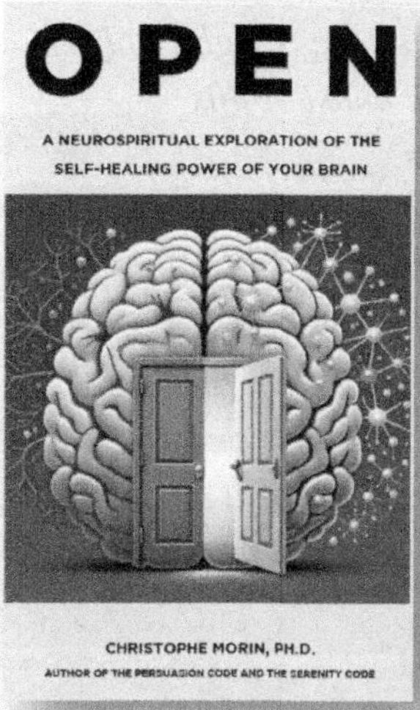

For as long as I can remember, I've been drawn to helping others, especially those in crisis. I felt a deep-seated need to fix things, to be the one who could solve problems that seemed too big for those around me. A constant willingness to step in, rescue, and save seemed noble. But beneath that appearance of selflessness was a darker truth I didn't want to face: I had built my identity around helping others because I didn't know how to help myself.

I was caught in what I now understand as the Savior Complex—a pattern so deeply rooted in my unconscious that I believed it defined me. It drove my relationships, shaped my career, and governed how I saw the world. And yet, despite my efforts to help, I often felt overwhelmed, frustrated, and, ultimately, empty. The more I gave, the more I was consumed by the needs of others and the less I understood my own. I didn't see it at the time, but my need to be the "savior" was trapping me in a cycle of stress, anxiety, and emotional turmoil.

It wasn't until I experienced a profound awakening that I could see the truth. I was trapped—not by the demands of those I helped, but by my unconscious need to save. This need had grown over time, largely unnoticed, into a core part of my identity. This realization, painful yet liberating, marked the beginning of my journey toward a life of openness and flow.

The Savior Complex is a powerful example of how unconscious patterns dictate our lives. Behaviors, beliefs, and emotions can trap us that we don't fully understand, yet we allow them to govern our actions. We do what we believe is right, only to find ourselves in pain. And it is here, in the midst of this struggle, that the potential for awakening lies.

Through this very personal book, I'll take you on my journey to break free from my unconscious patterns and embrace an OPEN life characterized by awareness, plasticity, and, most importantly, flow. The OPEN model is a neurospiritual guide to help you raise your consciousness, rewrite your story, and reprogram your life for deeper fulfillment and joy. But it all starts with one critical step: becoming aware of the unconscious forces that drive us.

Summary of Chapter 1: Living Fully OPEN

My journey to openness and growth didn't begin in a quiet therapy room or with a self-help book. It started in the depths of the Amazonian jungle, surrounded by the sounds of nature and the powerful presence of ancient plant medicines. In the embrace of the jungle, I truly began to understand what it meant to live an OPEN life.

The Amazon was a physical and metaphorical jungle for me—a place where my mind, body, and spirit confronted the unknown. I first visited Peru in 2015 and decided to return in 2023 to participate in a 26-day immersion into the ancient healing traditions of the Shipibo people. But this was more than just an exotic journey; it was a journey into the depths of my unconscious, a test of my ability to embrace experiences that stretched me beyond my comfort zone.

During this shamanic initiation retreat, I participated in multiple Ayahuasca ceremonies—each one peeling back layers of my mind, forcing me to confront parts of myself I had long ignored. The ancient plant medicine revealed the many ways my desire to control and save others had manifested my fear. In vivid detail, I saw how my Savior Complex had kept me trapped, always focusing outward instead of inward, always searching for meaning in the problems of others rather than within myself. But the jungle did more than show me my limitations—it showed me the path forward. It taught me that to live OPEN, I had to let go. I had to embrace uncertainty, lean into discomfort, and allow experiences to reshape me. In the jungle, I first felt the true power of neuroplasticity—the brain's incredible ability to rewire itself. The experience reminded me that we are not fixed in our ways of thinking or being. We can grow, change, and expand—if we embrace new experiences with an OPEN mind.

Living an OPEN life is about more than just self-awareness. It's about actively seeking growth through experience. Openness is the key to unlocking the mind's full potential through ancient healing practices, travel, or simply stepping into the unknown.

Summary of Chapter 2: The Self-healing Power of Consciousness and Neuroplasticity

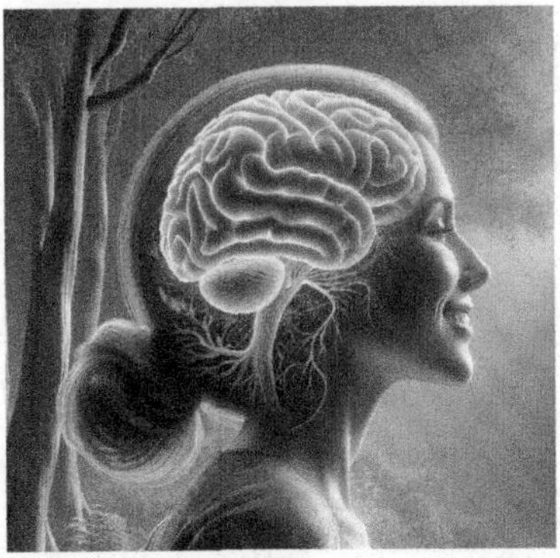

As a neuroscientist, I have discovered something truly transformative: the brain is not a static organ. Chapter 2 dives deeply into the transformative science of the brain, exploring neuroplasticity and emerging models from quantum mechanics, neurospirituality, and artificial intelligence to provide a holistic view of how we can unlock human potential. The chapter celebrates the brain's dynamic capacity to rewire itself, form new neural pathways, and adapt to challenges. Neuroplasticity, the brain's inherent ability to change, offers hope for profound personal transformation, allowing you to reshape thoughts, beliefs, and behaviors no matter their starting point.

This chapter goes beyond traditional neuroscience, introducing groundbreaking perspectives that merge science with spirituality. It discusses quantum consciousness models that propose the interconnectedness of the mind and universe, challenging the materialistic view of the brain as a closed system. These ideas intersect with other neurospiritual models, emphasizing how meditation, mindfulness, and even psychedelic experiences can reprogram deeply ingrained neural patterns to promote healing and growth. AI is also examined for its potential to advance our understanding of cognitive processes, offering parallels to the adaptability and learning capabilities of the human mind.

Ultimately, Chapter 2 paints a vision of the brain as a limitless organ, bridging science, spirituality, and technology. It invites readers to embark on an ongoing journey of self-discovery and transformation, embracing the potential of a mind liberated by neuroplasticity, inspired by quantum and spiritual insights, and enriched by modern innovation.

Summary of Chapter 3: OPEN Your Mind

Consciousness is the cornerstone of living in an OPEN flow state. But what does it truly mean to elevate your awareness? It involves moving beyond surface-level thinking to cultivate a profound connection with your thoughts, emotions, and surroundings. Expanding your consciousness allows you to approach life's challenges with greater clarity, presence, and purpose.

This chapter explores the transformative power of heightened awareness, offering practical tools to help you unlock this potential. Techniques such as mindfulness, meditation, and breathwork take center stage, providing accessible methods to deepen your connection to the present moment. These practices enhance your ability to focus and respond and create the conditions for entering a flow state—a mental zone where you feel fully engaged, alive, and attuned to life's rhythms.

A key focus is identifying and overcoming barriers to consciousness. Some obstacles, such as distractions or negative thought patterns, are evident, while others lie hidden in the subconscious, shaped by past experiences or unresolved fears. This chapter guides recognizing these deeper blockages and gently working through them. Raising your consciousness is a journey of continual growth. It opens the door to living with greater intention, enabling you to experience life more fully and authentically. By cultivating awareness and embracing the tools provided, you will discover a sense of flow that transforms how you engage with the world, fostering alignment, fulfillment, and peace.

Summary of Chapter 4: OPEN Your Self

We all live by stories we've been told and tell ourselves. But what if the stories that shape your life no longer serve you? What if the narratives running through your mind limit your growth and keep you stuck in patterns of stress, anxiety, or self-doubt? In this chapter, you'll learn how to reclaim your story by identifying the narratives that hold you back and rewriting them. You'll explore the neuroscience behind the Self and how changing the narrative in your mind can change your reality. You can create a new story that aligns with your true potential with the tools provided.

The chapter integrates two powerful personality frameworks to help you examine and transform your narratives: the Big Five Personality Traits (OCEAN) and the Enneagram. The Big Five traits—Openness, Conscientiousness, Extraversion, Agreeableness, and Neuroticism—offer a scientific lens to understand tendencies and behaviors. On the other hand, the Enneagram's nine types illuminate ego-driven survival strategies and pathways to transcendence. Together, these systems allow you to see how their unique traits and patterns interact with the stories they tell themselves.

Summary of Chapter 5: OPEN Your Life

Once you've elevated your awareness and rewritten your story, the next step is reprogramming your life. This chapter centers on embedding rituals that sustain an OPEN mind and maintain a flow state. It provides practical tools to reinforce positive mental pathways and prevent relapses into old, limiting patterns. Through PRIMAL, RATIONAL, and SPIRITUAL rituals, you can build a foundation for continuous growth and adaptability.

PRIMAL rituals include opening up to regular exercise, which boosts mood and cognition, and mindful eating and sleep hygiene, which promote energy and clarity. These habits create a strong mind-body connection that anchors you in the present. RATIONAL rituals focus on opening up to practices that promote mental resilience and self-awareness. Creative activities such as painting, traveling, playing, gardening, or cooking foster flow states, immersing you in experiences that can be challenging yet deeply rewarding for your neurons. SPIRITUAL rituals deepen your connection to purpose and meaning. Prayer, meditation, and other modalities align your actions with higher values.

The chapter emphasizes consistency and personalization, encouraging small, manageable steps to integrate these rituals seamlessly into your life. Over time, these practices cultivate resilience, openness, and a sense of joy. Reprogramming your life is not about achieving perfection but building a lifestyle aligned with your true potential. By embracing these rituals, you unlock a life of flow, adaptability, and fulfillment.

Summary of Conclusion: Flow into Your OPEN Future

Living an OPEN life is not a fixed destination but an ongoing, ever-evolving journey. It's about cultivating a mindset that sees each day as an opportunity for growth and expansion. As you progress, the focus shifts to sustaining the flow state you've created—where you feel aligned, energized, and deeply connected to life's possibilities.

To live in this state, embracing openness as a guiding principle is essential. This means remaining curious and receptive to new experiences, even those that challenge your comfort zone. Each moment invites you to learn, adapt, and deepen your understanding of yourself and the world. By approaching life with this mindset, you build resilience and cultivate a sense of wonder that fuels continuous transformation.

Trust in the power of your mind to guide you on this journey. Through the practices and insights you've gained, you've equipped yourself with the tools to navigate life's challenges with clarity and intention. Whether through mindful reflection, intentional rituals, or conscious action, each step reinforces your ability to remain in flow.

An OPEN life is a dynamic process—a balance of effort and surrender, action and reflection. By committing to this path, you unlock a more profound sense of fulfillment and joy, knowing that growth is always possible, and life's potential is boundless. Trust the process, and let each day guide you closer to your highest self.

INTRODUCTION

The Power of The Enneagram

OPEN UP is a transformative resource rooted in the principles of the *OPEN neurospiritual framework*, which blends neuroscience, neuroplasticity, and spiritual wisdom. This guide empowers you to harness the insights of the Enneagram—a typology of nine distinct personality types—as a diagnostic tool to uncover the unconscious coping mechanisms that drive stress, anxiety, depression, addiction, and trauma (SADAT).

By identifying your core fears, motivations, and habitual behaviors of their type, you will understand how these unconscious patterns may perpetuate suffering. Combined with the *OPEN* model, the Enneagram provides self-awareness and a pathway to reprogram these patterns through enhanced consciousness and strengthened neuroplasticity.

At its core, this guide is about transformation: the ability to transcend your ego's limitations, integrate shadow aspects, and embody the higher virtues associated with one's *Holy Idea.* The ultimate goal is accelerating growth and openness, enabling you to live with joy, purpose, and flow.

Be sure to take the complimentary *OPEN Enneagram test* before you dive into the book at https://su.vc/nfyybfft.

Understanding the Critical Sections for Each Type

The guide includes critical sections for each Enneagram type, offering a comprehensive framework to support healing and personal development. These sections help all readers diagnose and address the unconscious patterns contributing to Stress, Anxiety, Depression, Addictions, and Trauma (I created the acronym SADAT to reference all conditions) while cultivating their unique strengths and potential.

- **Core overview of each type**
  Each type is introduced with a clear summary of its defining characteristics, core motivations, and fears. This overview is a starting point for understanding how the types' perspective shapes thoughts, emotions, and behaviors.

- **Strengths and gifts of each type**
  This section highlights the innate talents and virtues of each type, such as courage, empathy, or innovation, highlighting how these qualities can create meaningful contributions in life and relationships.

- **Challenges and blind spots of each type**
  This is where I discuss the types' vulnerabilities and limitations, such as tendencies toward avoidance, rigidity, or excessive self-reliance, clarifying how these patterns may exacerbate suffering or hinder growth.

- **Key coping Mechanism of each type**
  An exploration of the unconscious defense strategies each type employs to navigate challenges. These mechanisms often evolve into maladaptive behaviors tied to SADAT conditions, such as overwork, conflict avoidance, or substance use.

- **Stress and growth paths for each type (Integration/Disintegration)**
  This is a detailed map of how the type behaves under stress and during periods of growth. By understanding these paths, readers can identify moments when they are regressing or evolving and take intentional steps toward positive transformation.

- **Wing influences of each type**
  Adjacent Enneagram types, or wings, provide nuance to personality expression. This section explains how wings shape individual behavior, adding flexibility and depth to self-understanding.

- **Triadic grouping (Gut, Heart, Head)**
  Understanding these group dynamics enhances awareness of the type's instinctual reactions and decision-making patterns. All 9 types break down according to the shared fears and emotional responses of each triadic group—Gut (8, 9, 1),

Heart (2, 3, 4), and Head (5, 6, 7).

- **Type-Specific recommended OPEN rituals**
  These practices foster emotional resilience, mindfulness, and openness, enabling sustainable growth. Personalized OPEN rituals—categorized as PRIMAL, RATIONAL, and SPIRITUAL—are provided to help each type address core vulnerabilities.

- **Holy Idea and virtue of each type**
  This section, a cornerstone of the guide, introduces the Holy Idea (the archetypal truth) and the virtue associated with each type. By transcending ego fixations, individuals can embody higher consciousness and connect with their true essence.

- **Temenos and shadow integration**
  This section introduces the concept of the temenos—a sacred psychological or physical space for self-exploration and integration. Here, individuals are guided to confront and reconcile shadow aspects, fostering deeper authenticity and self-acceptance.

- **Relationship dynamics between all types**
  This section offers actionable advice for improving communication, resolving conflicts, and fostering empathy. Get practical insights into how each type relates to others in professional, familial, and romantic contexts.

- **Type and SADAT framework**
  This section links personality insights with the likelihood of experiencing any SADAT conditions, offering personalized strategies to address emotional and psychological challenges. The type's unique traits are connected to the broader Neurospiritual Framework of the OPEN model.

Empowering Transformation Through a Neurospiritual Framework

OPEN UP goes beyond self-awareness, offering a comprehensive approach to healing and growth by integrating the Enneagram and the *OPEN neurospiritual framework.* By raising consciousness and enhancing neuroplasticity, the guide equips readers with the tools to transcend the unconscious patterns perpetuating SADAT conditions. This book invites you to transform life's challenges into opportunities for growth and fulfillment. By embracing the wisdom of the Enneagram and the principles of *OPEN*, you are empowered to live with clarity, connection, and boundless joy. With the help of *OPEN UP*, you can:

- Diagnose the extent to which unconscious behaviors contribute to SADAT conditions.
- Actively work on rewiring thought patterns through type-specific *OPEN rituals.*
- Nurture a more open, connected, and joyful life by integrating shadow aspects and embracing their higher potential.

The Power of the Enneagram Triads

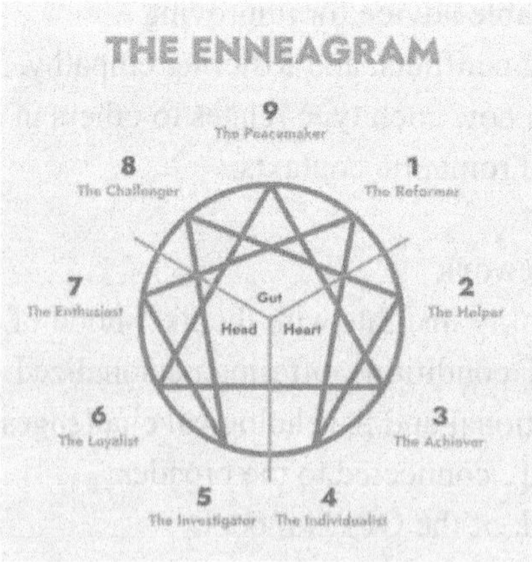

Figure 1: Enneagram symbol (Istock)

Like *OPEN*, the Enneagram model is a powerful system representing human consciousness. Understanding the more profound properties of Enneagram types—such as their triads, wings, and movement in stress and growth—is essential because no type exists in isolation. Triads (Gut, Heart, Head) reveal the core fears and emotional centers that drive each type's behavior. Recognizing these clusters of behaviors allows for a more fluid, neuroplastic approach to personal transformation rather than a rigid, fixed identity.

Meanwhile, The Enneagram symbol is a powerful representation of human consciousness, illustrating the interconnected nature of personality, growth, and transformation. It consists of a circle, a triangle (3-6-9), and a hexad (1-4-2-8-5-7), symbolizing the dynamic movement between different states of awareness. The circle represents wholeness and unity, emphasizing that all types are part of a greater whole. Unlike static personality models, the Enneagram is a living map of transformation, revealing the hidden patterns shaping our thoughts, emotions, and behaviors.

Here is the name and a short description of each type:

- Type 1: The Reformer – Analytical, idealistic, ethical, disciplined
- Type 2: The Helper – Compassionate, relational, giving, attentive
- Type 3: The Achiever – Goal-driven, ambitious, high-performing, image-focused
- Type 4: The Individualist – Emotional, introspective, creative, moody
- Type 5: The Investigator – Intellectual, observant, innovative, private
- Type 6: The Loyalist – Devoted, cautious, reliable, apprehensive
- Type 7: The Enthusiast – Energetic, adventurous, adaptable, distracted
- Type 8: The Challenger – Assertive, commanding, confident, forceful
- Type 9: The Peacemaker – Calm, accommodating, open-minded, passive

Enneagram types can be grouped into three sets based on shared core fears, reflecting deeper motivations and emotional responses. This grouping is not immediately apparent from the basic descriptions of each type, but it becomes more meaningful when considering their underlying motivations.

- *Types 8, 9, and 1 are GUT types and share the core fear of being vulnerable.* The core fears of the 8, 9, and 1 types relate to autonomy, control, and physical integrity. Each type copes with this fear differently: Type 8 asserts control, Type 9 avoids conflict to maintain control, and Type 1 self-controls to prevent being 'bad.'

- *Types 2, 3, and 4 are HEART types and share the core fear of being unworthy.* For types 2, 3, and 4, the central fear revolves around how others see and value them. Each type manages this fear differently: Type 2 gains love through helping, Type 3 through achievement, and Type 4 through individuality.

- *Types 5, 6, and 7 are HEAD types and fear the core fear of not knowing.* For these types, the core fear concerns security and coping with uncertainty. Type 5 manages this through detachment and self-reliance, Type 6 through loyalty and security, and Type 7 by escaping from pain and seeking variety

Integration and disintegration arrows for each type

Figure 1 features the arrows of integration (growth) and disintegration (stress). Each type evolves or struggles based on its circumstances. By recognizing these pathways, individuals can become more self-aware and intentional in their personal growth. When under stress, a type may adopt the less healthy traits of another number (disintegration). Conversely, when integrating and

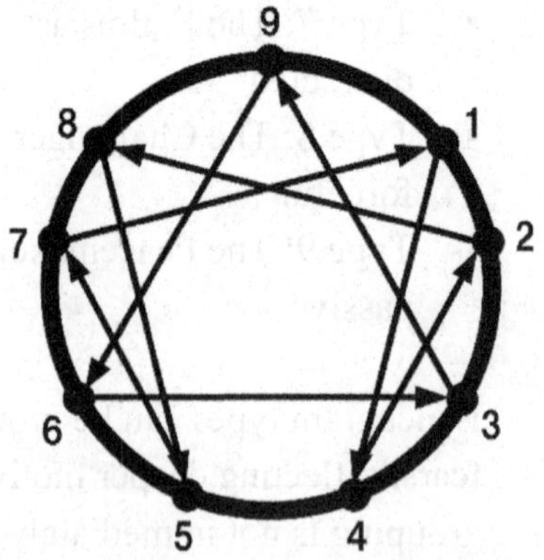

Figure 2: Enneagram Arrows (Istock)

growing, a type accesses the healthier traits of a different number (integration).

Type 1 (The Reformer or Perfectionist)

- **Under stress (Moves to 4):** Becomes moody, self-absorbed, and overly focused on their emotions. They may withdraw or feel that their ideals will never be achieved.
- **Integration (Moves to 7):** Becomes more spontaneous, playful, and optimistic, allowing themselves to enjoy life and embrace flexibility.

Type 2 (The Helper)

- **Under stress (Moves to 8):** Becomes aggressive, domineering, and demanding. They may use force to gain attention or feel resentful when their efforts aren't appreciated.
- **Integration (Moves to 4):** Becomes more self-aware, reflective, and authentic. They learn to focus on their own needs and express emotions deeply.

Type 3 (The Achiever)

- **Under stress (Moves to 9):** Disengaged, apathetic, and unmotivated. They may withdraw from productivity and avoid responsibility.
- **Integration (Moves to 6):** Becomes more cooperative, loyal, and oriented toward group well-being. They learn to value teamwork and authenticity over appearances.

Type 4 (The Individualist)

- **Under stress (Moves to 2):** Becomes overly people-pleasing, desperate for attention, and dependent on others' validation. They may sacrifice authenticity to gain love.
- **Integration (Moves to 1):** Becomes more disciplined, objective, and action-oriented. They channel their creativity into meaningful, constructive work.

Type 5 (The Investigator)

- **Under stress (Moves to 7):** Becomes scattered, impulsive, and distractible. They may overindulge in stimulation or avoid emotional discomfort through distractions.
- **Integration (Moves to 8):** Becomes more decisive, confident, and willing to act. They learn to trust their instincts and engage with the world more fully.

Type 6 (The Loyalist)

- **Under stress (Moves to 3):** Becomes overly image-conscious, competitive, and focused on gaining approval. They may overwork or obsess about success to feel secure.
- **Integration (Moves to 9):** Becomes more relaxed, trusting, and able to go with the flow. They develop inner peace and a balanced outlook on life.

Type 7 (The Enthusiast)

- **Under stress (Moves to 1):** Becomes overly critical, rigid, and perfectionistic. They may impose rules or discipline on themselves to feel in control.
- **Integration (Moves to 5):** Becomes more focused, reflective, and able to delve deeply into one thing at a time. They embrace depth over superficiality.

Type 8 (The Challenger)

- **Under stress (Moves to 5):** Becomes withdrawn, overly analytical, and detached. They may isolate themselves to avoid vulnerability or overanalyze their feelings.
- **Integration (Moves to 2):** Becomes more compassionate, nurturing, and open-hearted. They learn to show care and vulnerability in their relationships.

Type 9 (The Peacemaker)

- **Under stress (Moves to 6):** Becomes anxious, reactive, and overly focused on potential problems. They may struggle to make decisions or worry excessively.

- **Integration (Moves to 3):** Becomes more energetic, ambitious, and proactive. They take the initiative and express their individuality with confidence.

The role of wings for each type

Meanwhile, each Enneagram type can be influenced by one or both of its adjacent types, called *wings***.** These wings modify and enhance a type's personality, creating nuanced variations in how it expresses itself. While individuals may lean strongly toward one wing, both wings can play a role at different points in life.

Type 1 (The Reformer or Perfectionist)

- **Wing 9 (The Idealist):** Becomes more peaceful, balanced, and focused on maintaining harmony. This wing softens Type 1's rigidity, making them more patient and empathetic.
- **Wing 2 (The Advocate):** Becomes more nurturing, supportive, and people-oriented. This wing amplifies Type 1's desire to help others and emphasizes interpersonal connections.

Type 2 (The Helper)

- **Wing 1 (The Servant):** Becomes more principled, disciplined, and focused on doing what's "right." This wing adds structure and a moral compass to Type 2's giving nature.
- **Wing 3 (The Host/Hostess):** Type 2 becomes more ambitious, charming, and image-conscious. This wing encourages Type 2 to seek recognition for their generosity and social skills.

Type 3 (The Achiever)

- **Wing 2 (The Charmer):** Becomes more people-oriented, empathetic, and focused on helping others succeed. This wing softens Type 3's competitive edge with warmth and connection.
- **Wing 4 (The Professional):** The wing encourages Type 3 to become more reflective, creative, and focused on

authenticity. It also encourages Type 3 to align their ambitions with their deeper values.

Type 4 (The Individualist)

- **Wing 3 (The Aristocrat):** Becomes more driven, outwardly focused, and socially aware. This wing encourages Type 4 to channel their creativity into achievement and success.
- **Wing 5 (The Bohemian):** Becomes more intellectual, private, and focused on introspection. This wing adds a cerebral, philosophical quality to Type 4's emotional depth.

Type 5 (The Investigator)

- **Wing 4 (The Iconoclast):** Becomes more creative, emotionally attuned, and focused on self-expression. This wing enhances Type 5's connection to beauty and meaning.
- **Wing 6 (The Problem-Solver):** Becomes more practical, loyal, and focused on security. This wing adds an element of teamwork and dependability to Type 5's independence.

Type 6 (The Loyalist)

- **Wing 5 (The Defender):** Becomes more intellectual, independent, and focused on preparation. This wing strengthens Type 6's ability to analyze and create security through knowledge.
- **Wing 7 (The Buddy):** Becomes more sociable, playful, and adventurous. This wing adds lightheartedness and optimism to Type 6's cautious nature.

Type 7 (The Enthusiast)

- **Wing 6 (The Entertainer):** Becomes more grounded, loyal, and focused on building relationships. This wing balances Type 7's impulsive energy with practicality and trustworthiness.
- **Wing 8 (The Realist):** Becomes more assertive, confident, and focused on taking control. This wing amplifies Type 7's boldness and ability to take decisive action.

Type 8 (The Challenger)

- **Wing 7 (The Maverick):** Becomes more adventurous, playful, and creative. This wing lightens Type 8's intensity with spontaneity and a sense of fun.
- **Wing 9 (The Bear):** Becomes more grounded, calm, and focused on maintaining harmony. This wing softens Type 8's confrontational tendencies with patience and empathy.

Type 9 (The Peacemaker)

- **Wing 8 (The Referee):** Becomes more assertive, decisive, and willing to engage in conflict when necessary. This wing adds strength and leadership to Type 9's peacekeeping nature.
- **Wing 1 (The Dreamer):** Becomes more idealistic, disciplined, and focused on improvement. This wing enhances Type 9's desire to make the world a better place.

Key Insights About Wings

Understanding wings helps deepen self-awareness and explains variations within each type. It provides a richer understanding of how individuals express their Enneagram type daily. However, remember that:

- Wings are not fixed and can shift depending on circumstances or personal growth.
- Balancing both wings can help you access a fuller range of traits and behaviors.
- Over-reliance on one wing can create imbalance, amplifying that wing's challenges.

THE GUT TRIAD

GUT Types (8,9,1): Fear of Being Vulnerable

As noted, types 8, 9, and 1 are primarily concerned with control and resistance to control, relating to physicality and anger. Their core fear is to feel powerless, therefore exposed or vulnerable.

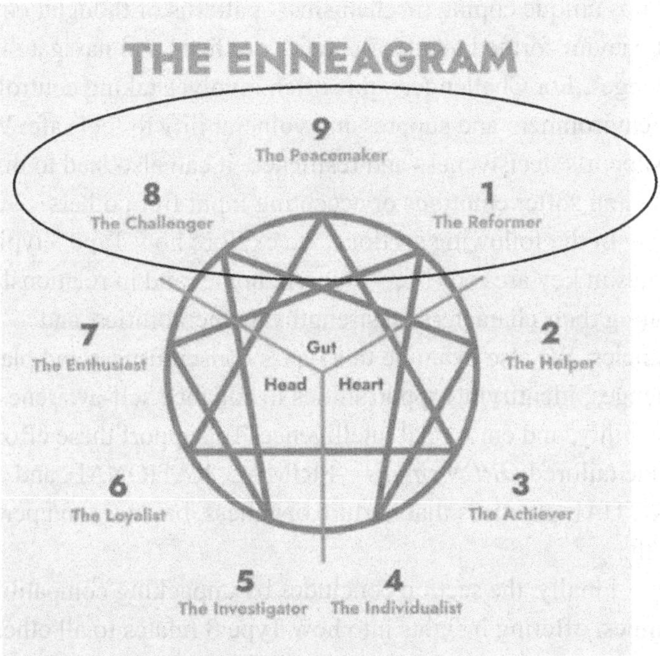

- **Type 8 (The Challenger):** Their core fear is to be controlled or harmed by others, which drives them to assert control over their environment and maintain independence at all costs.

- **Type 9 (The Peacemaker):** They fear loss and separation, leading to a desire for peace and stability. This is often a reaction to a perceived threat to their comfort and status quo, which they seek to maintain through avoidance of conflict.

- **Type 1 (The Perfectionist):** They fear being corrupt, evil, or defective. This fear drives their desire for order and integrity, often manifesting as internal control over impulses they deem wrong.

Type 8: The Challenger (Assertive, commanding, confident, forceful)

Confident and assertive, Challengers are natural leaders who value strength, independence, and control. They desire to protect themselves and others from harm or vulnerability. Each type develops unique coping mechanisms—patterns of thought, emotion, and behavior formed to address their core fears and navigate life's challenges. For Challengers, this often involves taking control of their environment and suppressing vulnerability to feel safe. While this supports decisiveness and resilience, it can also lead to difficulty expressing softer emotions or accepting input from others.

In the following sections, we explore how Type 8 typically operates in key areas of life—at work, home, and in relationships—revealing their characteristic strengths, vulnerabilities, and tendencies. We also examine this type's consciousness and plasticity challenges, identifying opportunities to enhance self-awareness, adaptability, and emotional intelligence. To support these efforts, we provide tailored *OPEN rituals*—PRIMAL, RATIONAL, and SPIRITUAL practices that nurture openness, balance, and personal growth.

Finally, the section concludes by unpacking compatibility dynamics, offering insights into how Type 8 relates to all other Enneagram types across various professional, familial, or romantic relationships.

The Gifts of Being an Enneagram 8

1. **Courage and Strength**: Type 8s are incredibly strong-willed and courageous. They face challenges head-on and are fearless in taking risks. One of their greatest strengths is their ability to stand firm in adversity.

2. **Protectiveness and Loyalty**: Type 8s fiercely protect their loved ones and those they feel responsible for. Their loyalty makes them reliable, trustworthy, and someone others can depend on.

3. **Decisiveness and Leadership**: Type 8's natural ability to take

charge and make confident decisions makes them excellent leaders. They are action-oriented and know how to get things done efficiently.

4. **Honesty and Integrity**: Type 8s value truth and transparency in all aspects of life. They speak their minds and encourage others to do the same, fostering environments of honesty and openness.

Type 8 Coping Mechanisms

As children, many Type 8s felt they had to protect themselves from vulnerability, often adopting coping mechanisms that emphasized strength, control, and independence. These mechanisms helped them survive and thrive in challenging environments and created specific patterns that can present challenges in adulthood.

1. **Fear of Vulnerability**

   - **Coping Mechanism**: In childhood, Type 8s may have learned that being vulnerable meant being hurt or taken advantage of, so they developed a tough exterior to protect themselves.
   - **Adult Challenge**: This fear of vulnerability can create emotional walls in relationships, making it difficult for them to fully open up and connect with others on a deeper level.

2. **Need for Control**

   - **Coping Mechanism**: Growing up in environments where they felt unsafe or powerless, Type 8s developed a need for control to avoid being manipulated or harmed.
   - **Adult Challenge**: This need for control can lead to overbearing or dominating behavior, causing tension in relationships and work environments where collaboration and shared decision-making are necessary.

3. **Suppressing Soft Emotions**

- **Coping Mechanism**: As children, Type 8s may have learned to suppress emotions like fear, sadness, or tenderness to avoid being seen as weak. Instead, they channeled their feelings into strength and anger, which felt safer.
- **Adult Challenge**: This emotional suppression can lead to difficulties in expressing softer emotions in adulthood, creating distance in relationships where vulnerability and empathy are essential.

4. **Resistance to Authority**

- **Coping Mechanism**: Many Type 8s felt they could not rely on authority figures in childhood, leading them to become highly self-reliant and resistant to being controlled by others.
- **Adult Challenge**: In adulthood, this can manifest as resistance to collaboration or challenges with authority figures, making it difficult to work in hierarchical environments or trust others' leadership.

Being an Enneagram Type 8 has incredible gifts, including courage, loyalty, strength, and honesty. They are natural leaders, protectors, and advocates who bring a sense of power and justice into every area of their lives. However, many of their strengths come from coping mechanisms adopted in childhood, such as the need to avoid vulnerability, control their environment, and suppress emotions.

These patterns, while protective in childhood, can create challenges in adulthood, particularly in relationships and leadership roles. To fully embrace their gifts, Type 8s must soften defenses, allow for vulnerability, and trust others. By learning to let go of some control and opening emotionally, Type 8s can deepen their relationships and become even more effective leaders while maintaining the strength and integrity that makes them such powerful forces in the world.

Type 8 at Work

Strengths

⇒ **Leadership and Authority**: Type 8s thrive in leadership roles. They are decisive, confident, and not afraid to take charge. They naturally gravitate toward positions where they can influence outcomes and protect others.

⇒ **Protective of Team Members**: They are fiercely loyal and protective of their team or colleagues, often standing up for those who cannot stand up for themselves. This makes them great advocates in the workplace.

⇒ **Action-oriented**: Type 8s are focused on achieving results. They don't shy away from challenges or difficult tasks, making them effective problem-solvers and drivers of change.

Challenges

⇒ **Overbearing Leadership**: Their desire for control and authority can be domineering, especially when they don't trust others to take initiative. This can lead to tension or power struggles with colleagues or subordinates.

⇒ **Impatience with Weakness**: Type 8s may have little patience for perceived weakness or indecision in others, leading to frustration and conflict.

⇒ **Difficulty Delegating**: Because they trust their abilities over others, Type 8s may struggle to delegate tasks, which can overwhelm them or alienate coworkers who want more responsibility.

Type 8 at Home

Strengths

⇒ **Protective of Family**: Type 8s often take on the role of protector in the family, ensuring their loved ones are safe and cared for. They are dependable and strong, providing a sense of security.

⇒ **Direct and Honest**: They are forthright in their communication and expect the same in return. While they might not sugarcoat things, their honesty fosters clear communication and trust.

⇒ **Creating Stability**: Type 8s bring stability and structure to the household. They enjoy taking on responsibilities and ensuring the family operates smoothly, especially in times of crisis.

Challenges

⇒ **Control Issues**: Their need for control at home can lead to conflicts with family members, especially if others feel they are being micromanaged or overpowered.

⇒ **Difficulty Showing Vulnerability**: Type 8s often suppress their vulnerability, believing it to be a weakness. This can make it hard for them to connect emotionally with family members, leading to distance in relationships.

⇒ **Power Struggles**: Because they are so used to being in control, Type 8s may engage in power struggles with loved ones, particularly if they feel challenged or disrespected.

Type 8 in Relationship

Strengths

⇒ **Loyalty and Commitment**: Once Type 8s commit to a relationship, they are deeply loyal and protective. They take their commitments seriously and are highly dependable.

⇒ **Confidence**: Type 8s' confidence can attract partners, as they bring a sense of stability and assertiveness to relationships. They often take charge and provide direction, which can reassure their partner.

⇒ **Honesty and Integrity**: They value honesty in relationships and are very direct with their thoughts and feelings. They expect the same level of transparency from their partner.

Challenges

⇒ **Fear of Vulnerability**: Type 8s often struggle to express softer emotions, such as fear or hurt, because they fear being seen as weak. This can make it difficult for them to fully open up and be emotionally vulnerable in relationships.

⇒ **Controlling Nature**: Their desire for control can create an imbalance in relationships, where their partner may feel dominated or undervalued.

⇒ **Impatience and Intensity**: Type 8s bring an intense energy to relationships. This intensity can sometimes feel overwhelming for partners who prefer a more relaxed or gentle approach.

Type 8 Consciousness Challenges

For Enneagram Type 8s, the core challenges related to consciousness and plasticity stem from their deeply ingrained need for control, protection from vulnerability, and their tendency to avoid showing weakness. These characteristics, while protective, can hinder their ability to grow emotionally, increase flexibility, and raise consciousness. Let's explore the specific challenges Type 8s face in these areas.

1. **Fear of Vulnerability**

Type 8s often suppress vulnerability by equating it with weakness. This makes it difficult for them to access and become aware of their more tender emotions, such as fear, sadness, or compassion. Their fear of vulnerability limits their emotional self-awareness. They might disconnect from their feelings or fail to understand the impact of their intense energy on others. By rejecting vulnerability, they miss the opportunity to fully experience and appreciate their inner emotional world, which is crucial for deeper consciousness.

2. **Control-Oriented Thinking**

Type 8s are driven by a need to control their environment and protect themselves from feeling powerless. While this gives them a sense of security, it can also narrow their perspective and reduce their openness to alternative viewpoints or unexpected outcomes. Their focus on control can create rigidity in thinking, which prevents them from being present and open to the flow of life. It can also make surrendering to what is beyond their control difficult— an essential part of raising consciousness and finding inner peace.

3. **Suppression of Emotions**

Because they view emotions like fear or sadness as weaknesses, Type 8s often suppress or redirect them into anger or action. This emotional suppression makes it difficult for them to process and understand their feelings. Emotional suppression lowers Type 8's emotional intelligence and awareness capacity. It prevents them from achieving emotional balance and hinders their ability to connect deeply with others.

4. **Over-identification with Strength**

Type 8s tend to over-identify with their external strength, using power and assertiveness as defenses against being hurt or manipulated. This creates an inflated sense of invulnerability. Over-identification with strength keeps Type 8s in a surface-level awareness that focuses on external power rather than internal truths. This limits their ability to access higher consciousness, which requires embracing vulnerability, introspection, and self-compassion.

Type 8 Plasticity Challenges

Neuroplasticity refers to the brain's ability to reorganize itself by forming new neural connections throughout life. It allows people to change habits, beliefs, and emotional responses through new experiences and learning. Type 8's strong tendencies toward control, rigidity, and self-reliance can limit their brain's ability to adapt, grow, and remain flexible.

Key Challenges:

1. **Need for Control**

Type 8s often want to control outcomes and environments to avoid feeling powerless. This habit keeps them in familiar patterns where they feel secure. The need for control reduces mental flexibility, as Type 8s may resist uncertainty and change, which are crucial for developing new neural pathways. Instead of adapting, they may double down on their existing beliefs and strategies, limiting their ability to learn from new experiences or open themselves up to different approaches.

2. **Rigidity and Resistance to Change**

Type 8s can resist external authority or anything threatening their independence. This resistance to change or submission to external influence can create rigidity in their thinking and actions. This resistance reduces the brain's ability to adapt to new circumstances or perspectives. By rejecting new ideas or approaches that aren't aligned with their own, Type 8s hinder their capacity for cognitive flexibility and slow down the creation of new neural pathways, which are vital for growth.

3. **Black-and-White Thinking**

Type 8s often think in terms of black-and-white, right or wrong, with little room for ambiguity. This simplified view of the world makes them decisive and limits their ability to see nuance or compromise. Black-and-white thinking reduces neuroplasticity because the brain is not being exercised to think in complex or nuanced ways. Embracing ambiguity, gray areas, or multiple perspectives is crucial for building more adaptable and flexible neural networks.

4. **Suppressing Emotional Learning**

Type 8s tend to suppress emotions that they deem "weak" and only express emotions aligned with strength (such as anger). This emotional suppression blocks them from learning from their more vulnerable feelings. By avoiding emotional vulnerability, type 8s miss out on the brain's ability to rewire emotional responses through emotional awareness and processing. Emotional experiences are essential for developing new neural connections that support emotional intelligence, empathy, and resilience.

5. **Over-reliance on Autonomy**

Type 8s are self-reliant and prefer to manage everything independently. While this fosters independence, it also isolates them from learning through collaboration, compromise, or feedback. Autonomy without openness to others' input limits social plasticity—the brain's ability to learn from social interactions, shared experiences, and collaboration. Relying solely on oneself prevents Type 8s from accessing new ways of thinking or problem-solving from working with others.

Type 8: Summary of Challenges

Challenge	Impact on Consciousness	Impact on Plasticity
Fear of Vulnerability	Limits emotional awareness and self-reflection.	Prevents emotional learning and rewiring emotional responses.
Need for Control	Hinders openness and surrender to the present moment.	Reduces mental flexibility and the ability to adapt to new inputs.
Emotional Suppression	Blocks connection to deeper emotions and emotional intelligence.	Limits neural connections associated with emotional regulation.
Rigidity and Resistance to Change	Closes off opportunities for growth and new perspectives.	Stifles cognitive flexibility and adaptability.
Black-and-White Thinking	It prevents nuanced understanding and limits higher consciousness.	Hinders complex thinking and the development of new neural pathways.
Over-reliance on Autonomy	Reduces the ability to learn through collaboration and trust in others	Limits social plasticity and the ability to learn from others

Type 8: PRIMAL Rituals Recommended

These rituals help Type 8s regulate their intense energy, manage stress, and maintain emotional control by engaging the body and grounding their nervous system.

Open up to Running

⇒ **Why**: Running is an excellent outlet for Type 8's natural intensity. It lowers cortisol, releases endorphins (which elevate mood), and reduces stress.

⇒ **Frequency**: Several times a week.

Open up to your Breath

⇒ **Why**: Breathwork can reduce negative emotions, calm the amygdala (the emotional center), and improve the body's vagal tone, helping Type 8s control their impulses.

⇒ **Frequency**: Daily for 5-10 minutes.

Open up to Nature

⇒ **Why**: Spending time in nature helps Type 8s reduce stress and re-establish a sense of calm, lowering blood pressure and strengthening the immune system.

⇒ **Frequency**: Weekly.

Open up to Your Sleep

⇒ **Why**: Prioritizing sleep helps Type 8s recharge their emotional reserves, aiding in emotional stability and reducing stress and anxiety.

⇒ **Frequency**: Every night, focusing on establishing a routine for better sleep quality.

Type 8: RATIONAL Rituals Recommended

RATIONAL rituals develop mental flexibility, shift black-and-white thinking, and open up Type 8s to new perspectives by engaging in activities that promote cognitive and emotional growth.

Open up to Laughter

⇒ **Why**: Humor reduces stress, increases the release of

endorphins, and helps Type 8s relax their intense, controlling tendencies.

⇒ **Frequency**: Weekly or as needed (e.g., watching comedies or engaging in fun activities).

Open up to Learning

⇒ **Why**: Challenging their minds through learning helps Type 8s improve cognitive flexibility and adaptability, encouraging them to embrace new perspectives.

⇒ **Frequency**: Weekly (e.g., reading, taking classes, or exploring new topics).

Open up to Cooking

⇒ **Why**: Cooking engages multiple senses and improves cognitive and emotional regulation. It also allows Type 8s to exercise creativity in a controlled yet enjoyable way.

⇒ **Frequency**: Weekly.

Open up to Gardening

⇒ **Why**: Gardening helps regulate emotions, reduces stress, and promotes cognitive flexibility by engaging Type 8s in a mindful, nurturing activity.

⇒ **Frequency**: Weekly or bi-weekly.

Type 8: SPIRITUAL Rituals Recommended

These rituals soften the desire for control, embrace vulnerability, and develop a deeper connection to their emotional and spiritual self. They encourage emotional vulnerability and reduce the need for control through spiritual practices that release ego-driven patterns.

Open up to Meditation

⇒ **Why**: Meditation helps Type 8s quiet the mind, regulate emotions, and reduce stress. It increases gray matter and fosters emotional vulnerability.

⇒ **Frequency**: Daily, for 10-15 minutes.

Open up to Stargazing

⇒ **Why**: Contemplating the vastness of the universe helps Type 8s shift their perspective, reducing their need to dominate and allowing them to connect with a sense of humility.

⇒ **Frequency**: Monthly or as needed to foster relaxation and wonder.

Open up to Mantras

⇒ **Why**: Repeating mantras improves focus, reduces stress, and helps Type 8s engage in self-reflection, thus reducing ego and increasing mindfulness.

⇒ **Frequency**: Daily, during meditation or quiet reflection.

Type 8: Compatibility With Other Types

Type 1 (The Reformer)

•**What Works**: Both value justice and integrity. Type 1's ethical drive pairs well with Type 8's desire for strength and leadership. Type 8 admires Type 1's principles and discipline, while Type 1 respects Type 8's decisiveness.
•**Potential Friction**: Type 1's rigidity and perfectionism may frustrate Type 8, who values autonomy and dislikes being micromanaged. Type 1 may see Type 8's forcefulness as reckless or overly aggressive.

Type 2 (The Helper)

•**What Works**: Type 2's nurturing and supportive nature complements Type 8's protective instincts. Type 8 provides strength, and Type 2 offers warmth and care, creating a balanced dynamic.
•**Potential Friction**: Type 8's independence may conflict with Type

2's need for emotional closeness and validation. If Type 8 becomes too overbearing, Type 2 may feel unappreciated, while Type 8 may see Type 2's emotional needs as clingy.

Type 3 (The Achiever)

•**What Works**: Both are ambitious and results-driven. Type 8's energy and decisiveness align well with Type 3's focus on success and achievement. Together, they can form a powerful, goal-oriented team.
•**Potential Friction**: Competition for control and recognition can arise. Type 8 may view Type 3's image-consciousness as superficial, while Type 3 might see Type 8's directness as too blunt or aggressive.

Type 4 (The Individualist)

•**What Works**: Type 4's authenticity and emotional depth may intrigue Type 8, while Type 4 can admire Type 8's strength and assertiveness. Both value being true to themselves.
•**Potential Friction**: Type 8's confrontational and action-oriented nature may clash with Type 4's introspection and emotional sensitivity. Type 8 might dismiss Type 4's feelings, while Type 4 may feel overwhelmed by Type 8's intensity.

Type 5 (The Investigator)

•**What Works**: Both types value independence and self-sufficiency. Type 8 appreciates Type 5's intellectual depth, and Type 5 respects Type 8's strength and decisiveness. Together, they create a dynamic balance of action and thought.
•**Potential Friction**: Type 8's need for control and engagement can overwhelm Type 5, who requires space and autonomy. Type 5's emotional detachment may frustrate Type 8, who values direct and intense interactions.

Type 6 (The Loyalist)

•**What Works**: Type 6's loyalty and dedication align well with Type 8's protective instincts. Type 8 offers the security and decisiveness Type 6 seeks, and Type 6 provides the commitment and reliability that Type 8 values.
•**Potential Friction**: Type 6's anxiety and indecisiveness may irritate Type 8, who prefers immediate action and clear direction. Conversely, Type 6 may feel bulldozed by Type 8's intensity and desire for control.

Type 7 (The Enthusiast)

•**What Works**: Both are energetic and love taking on challenges. Type 7's spontaneity complements Type 8's decisiveness; they can enjoy new experiences enthusiastically and excitedly.
•**Potential Friction**: Type 8 may find Type 7 too scattered or avoidant of serious matters, while Type 7 could feel restricted by Type 8's controlling or aggressive tendencies.

Type 8 (The Challenger)

•**What Works**: Two Type 8s can form a powerful, assertive partnership full of energy and decisiveness. They understand each other's need for control and action.
•**Potential Friction**: Power struggles are likely. Both are strong-willed and dominant, leading to potential conflicts over who takes the lead or makes the final decision. Mutual respect and compromise are essential for harmony.

Type 9 (The Peacemaker)

•**What Works**: Type 9's calm, easygoing nature can soften Type 8's intensity. Type 8 appreciates Type 9's non-confrontational support, while Type 9 feels protected by Type 8's strong leadership.
•**Potential Friction**: Type 9's tendency to avoid conflict and decision-making can frustrate Type 8, who values directness and resolution. Type 8 may see Type 9's passivity as a weakness, while

Type 9 may feel overwhelmed by Type 8's forcefulness.

In summary, **Type 8** pairs well with types that offer support, loyalty, or ambition but may clash with too passive, sensitive, or indecisive types. Successful relationships require mutual respect and understanding of Type 8's assertive and intense nature.

Type 8: Compatibility Table

Pair	Compatibility with Type 8	Strengths	Challenges	Tips for Improving Flow
8 and 1	Shared focus on justice and integrity can drive meaningful action.	Type 1's discipline complements Type 8's decisiveness.	Type 1's rigidity may clash with Type 8's intensity and need for control.	Type 8: Practice openness to Type 1's structured perspective to build mental flexibility. Type 1: Embrace Type 8's dynamic approach to challenge perfectionism.

Pair	Compatibility with Type 8	Strengths	Challenges	Tips for Improving Flow
8 and 2	Loyalty and protectiveness create a strong partnership for mutual growth.	Type 2's emotional support helps Type 8 explore vulnerability.	Type 8 may dismiss Type 2's emotional needs, while Type 2 may become overly accommodating.	Type 8: Raise consciousness by acknowledging emotions as strength. Type 2: Use plasticity habits like setting boundaries to avoid overgiving.
8 and 3	Ambition and drive make them a high-performing and results-oriented team.	Shared focus on success fosters synergy and mutual respect.	Competitive tendencies may arise; Type 8 may find Type 3 too focused on appearances.	Type 8: Enhance plasticity by exploring authenticity beyond achievements. Type 3: Expand consciousness by valuing Type 8's direct honesty.

Pair	Compatibility with Type 8	Strengths	Challenges	Tips for Improving Flow
8 and 4	Deep emotional connections are possible if both embrace vulnerability.	Type 4's creativity inspires Type 8 to explore emotional depth.	Type 8 may struggle with Type 4's sensitivity; Type 4 may feel overshadowed by Type 8's assertiveness.	Type 8: Increase plasticity by allowing emotional exploration. Type 4: Expand consciousness by seeing Type 8's strength as protective, not domineering.
8 and 5	Mutual respect for autonomy and intellectual depth fosters a balanced dynamic.	Type 5's analytical thinking complements Type 8's action-oriented nature.	Type 8 may overwhelm Type 5 with intensity; Type 5 may withdraw emotionally.	Type 8: Build consciousness by respecting Type 5's need for space. Type 5: Develop plasticity by engaging in collaborative decision-making with Type 8.

Pair	Compatibility with Type 8	Strengths	Challenges	Tips for Improving Flow
8 and 6	Shared loyalty and a protective nature create a secure partnership.	Type 6's cautious approach balances Type 8's boldness, fostering stability.	Type 8 may find Type 6 overly anxious; Type 6 may feel intimidated by Type 8's assertiveness.	Type 8: Raise consciousness by providing reassurance and fostering trust. Type 6: Strengthen plasticity by embracing risk-taking under Type 8's guidance.
8 and 7	High-energy pairings where spontaneity and boldness complement each other.	Type 7's optimism lightens Type 8's intensity, fostering creativity and action.	Type 8 may find Type 7 scattered; Type 7 may feel constrained by Type 8's control.	Type 8: Enhance plasticity by allowing spontaneity and exploring new experiences. Type 7: Expand consciousness by recognizing the value of focus and structure.

Pair	Compatibility with Type 8	Strengths	Challenges	Tips for Improving Flow
8 and 8	A powerful duo where mutual respect and shared intensity create dynamic energy.	Shared strength and confidence foster passionate and transformative growth.	Power struggles are common; both may resist vulnerability or flexibility.	Both: Practice consciousness through active vulnerability and shared reflection. Use plasticity habits like collaborative problem-solving to build trust.
8 and 9	Type 9's calmness balances Type 8's intensity, creating a harmonious relationship.	Type 9's adaptability and harmony-seeking nurture Type 8's emotional stability.	Type 8 may find Type 9 passive; Type 9 may withdraw from Type 8's assertiveness.	Type 8: Build consciousness by practicing patience and listening. Type 9: Develop plasticity by asserting needs and taking decisive actions with Type 8.

Type 9: The Peacemaker (Calm, accommodating, open-minded, passive)

Calm and accommodating, Peacemakers are driven by a desire to maintain harmony and avoid conflict. They bring a sense of stability and inclusiveness to their environments. Each type develops unique coping mechanisms—patterns of thought, emotion, and behavior formed to address their core fears and navigate life's challenges. For Peacemakers, this often involves avoiding confrontation and prioritizing the needs of others to maintain peace. While this fosters adaptability and diplomacy, it can lead to passivity or difficulty asserting themselves.

In the following sections, we explore how Type 9 typically operates in key areas of life—at work, home, and in relationships—revealing their characteristic strengths, vulnerabilities, and tendencies. We also examine this type's consciousness and plasticity challenges, identifying opportunities to enhance self-awareness, adaptability, and emotional intelligence. To support these efforts, we provide tailored *OPEN rituals*—PRIMAL, RATIONAL, and SPIRITUAL practices that nurture openness, balance, and personal growth.

Finally, the section concludes by unpacking compatibility dynamics, offering insights into how Type 9 relates to all other Enneagram types across various professional, familial, or romantic relationships.

The Gifts of Being an Enneagram 9

1. **Calming Presence**: Type 9s bring calming energy to every environment they enter, whether at work, home, or in relationships. Their ability to maintain peace and emotional steadiness is a gift to those around them.

2. **Empathy and Understanding**: Nines are naturally empathetic and excellent at understanding different perspectives. They often help others feel heard and valued, creating an atmosphere of acceptance and safety.

3. **Non-Judgmental Nature**: Nines are not quick to judge others. They accept people as they are, which makes them approachable and easy to confide in. This lack of judgment fosters strong, trusting relationships.

4. **Ability to Mediate and Harmonize**: Nines are skilled at seeing both sides of an argument and finding a middle ground. They make excellent mediators and have a gift for smoothing conflicts and harmonizing people together.

5. **Loyal and Supportive**: Type 9s are incredibly loyal and supportive once committed. They are dedicated to maintaining stability and balance in their relationships and are dependable partners, friends, and family members.

Type 9 Coping Mechanisms

Many of the challenges Type 9s face as adults are rooted in coping mechanisms developed in childhood. These mechanisms helped them survive and thrive in challenging environments, but they also created certain patterns that can present challenges in adulthood. These mechanisms were likely formed to maintain peace and avoid conflict or discomfort.

1. **Avoiding Conflict**

   - **Coping Mechanism**: As children, Type 9s may have learned to avoid conflict to keep the peace at home or in other settings. This avoidance became a way to protect themselves from emotional discomfort or confrontation.
   - **Adult Challenge**: In adulthood, this coping mechanism often leads to avoidance of difficult conversations, which can cause problems to fester. Nines may struggle to assert their needs or confront issues, leading to passive-aggressive behaviors or resentment.

2. **Merging with Others**

- **Coping Mechanism**: Type 9s may have learned to focus on others' needs and blend in with their surroundings to avoid being seen as a source of conflict. By "merging" with others' desires, they could maintain harmony.
- **Adult Challenge**: As adults, this leads to a loss of personal identity and difficulty asserting their own opinions, desires, and needs. Nines may feel invisible or undervalued in relationships because they prioritize others over themselves.

3. **Suppressing Personal Desires**

- **Coping Mechanism**: Type 9s may have learned to suppress their desires and feelings in childhood to keep things peaceful. They avoided making waves by putting their own needs last.
- **Adult Challenge**: As adults, this results in neglecting their goals and aspirations. Nines may struggle with self-motivation or feel disconnected from their true desires, leading to inertia or procrastination.

4. **Complacency and Inertia**

- **Coping Mechanism**: To maintain peace, Nines often found comfort in complacency, avoiding situations that required effort or might disrupt their inner calm.
- **Adult Challenge**: This leads to procrastination, lack of motivation, and an avoidance of challenges. Nines may struggle to take initiative or assert themselves in important decisions in their personal and professional lives.

While Enneagram Type 9s bring peace, harmony, empathy, and loyalty to their environments, their childhood coping mechanisms often lead to assertiveness, self-expression, and initiative challenges. The strategies they developed to avoid conflict and maintain peace, such as merging with others, suppressing their needs, and avoiding confrontation, can cause difficulties in adult relationships and personal growth. However, by recognizing these patterns and consciously working to assert their voice and desires,

Type 9s can maintain their natural gifts while cultivating a stronger sense of self, leading to more balanced and fulfilling lives.

Type 9 at Work

Strengths

⇒ **Calm Under Pressure**: Type 9s bring a sense of calm and steadiness to the workplace. They can diffuse tensions and help create a harmonious work environment by staying composed and even-tempered.

⇒ **Collaborative and Team-Oriented**: Nines prefer collaboration over competition and work well in teams. They listen to different viewpoints and are willing to accommodate others' needs to maintain a cooperative atmosphere.

⇒ **Adaptable and Easygoing**: They are highly adaptable and go with the flow, often content to follow others' lead rather than pushing their agenda. This flexibility makes them reliable team members who are easy to work with.

⇒ **Diplomatic and Mediating**: Nines have a talent for seeing different sides of an issue, making them effective mediators who can help resolve conflicts between colleagues. Their neutrality allows them to keep things balanced.

Challenges

⇒ **Avoidance of Conflict**: In their desire to maintain peace, Type 9s may avoid addressing problems or conflicts at work, leading to unresolved issues. This avoidance can result in them suppressing their opinions or needs, leading to resentment or passive-aggressive behavior.

⇒ **Lack of Assertiveness**: Nines may struggle to assert themselves, especially if they fear that doing so will disrupt the harmony they value. They may go along with others' ideas or decisions even when disagreeing, making it hard for their authentic voice to be heard.

⇒ **Procrastination and Inertia**: Type 9s can sometimes

become complacent or avoid taking action on essential tasks, especially if they feel overwhelmed. Their need for comfort and avoidance of discomfort can lead to procrastination.

Type 9 at Home

Strengths

⇒ **Nurturing and Harmonious**: At home, Type 9s nurture and create a peaceful and supportive environment. They are attentive to others' needs and go out of their way to ensure family members feel comfortable and secure.

⇒ **Supportive and Accommodating**: Nines are naturally supportive, tend to smooth over disagreements, and maintain balance in the family dynamic. They avoid drama and prefer a steady, stable home life.

⇒ **Emotionally Steady**: Type 9s are typically calm and unflappable, providing emotional steadiness for their family members. Their temperament makes them easy to live with, and they are usually content with a quiet, routine home life.

Challenges

⇒ **Neglecting Their Own Needs**: In their effort to keep the peace, Nines can neglect their own needs and desires. They may lose themselves while keeping everyone else happy, and this self-neglect can lead to feeling overlooked or undervalued.

⇒ **Difficulty Making Decisions**: Type 9s often struggle with decision-making, particularly when they fear their choices might cause conflict or upset. This indecision can frustrate family members and slow down essential processes at home.

⇒ **Passive Resistance**: Nines may resist change or challenges passively. Rather than directly expressing dissatisfaction, they may procrastinate or silently withdraw, creating tension in the household without anyone realizing why.

Type 9 in Relationship

Strengths

⇒ **Patient and Non-Judgmental**: In relationships, Type 9s are accepting, patient, and willing to see things from their partner's perspective. They rarely impose their own will and prefer to go with the flow, making them adaptable partners.

⇒ **Stable and Loyal**: Nines are intensely loyal and committed once they are in a relationship. Their steady presence provides a sense of security and trustworthiness. They are usually in it for the long haul and value the relationship's stability.

⇒ **Empathetic and Supportive**: Nines are good listeners and are naturally empathetic toward their partner's feelings. They want their loved ones to feel understood and supported, and they will go to great lengths to maintain harmony in the relationship.

Challenges

⇒ **Tendency to Merge with Partners**: Type 9s can lose their sense of relationship identity by focusing too much on their partner's needs and desires. This "merging" behavior can cause them to disconnect from their wants, leading to feelings of dissatisfaction or invisibility.

⇒ **Avoiding Difficult Conversations**: Nines are uncomfortable with confrontation and may avoid discussing relationship issues that need attention. Their reluctance to engage in difficult conversations can lead to unresolved problems and resentment over time.

⇒ **Passivity and Lack of Initiative**: In their desire to maintain peace, Nines may become passive in relationships, waiting for their partner to make decisions or initiate essential discussions. This lack of initiative can frustrate partners seeking more engagement or leadership from their 9 partners.

Type 9 Consciousness Challenges

Enneagram Type 9's key challenges related to consciousness and neuroplasticity arise from their tendencies toward avoidance of conflict, self-neglect, and passivity. Type 9s desire inner and outer peace and seek to maintain harmony in their environment. While these traits can bring stability, they can also limit emotional awareness, growth, and adaptability.

1. **Avoidance of Conflict and Discomfort**

Type 9s habitually avoid conflict to maintain harmony and inner peace. They tend to downplay issues, sidestep difficult conversations, or ignore problems rather than face them directly. Avoiding conflict stifles emotional consciousness. By avoiding discomfort, Type 9s often lose touch with their true feelings, wants, and needs, which leads to a lack of emotional clarity. They may struggle to recognize their emotional states or understand their internal motivations, making it harder to grow in self-awareness.

2. **Merging with Others**

Type 9s often merge with the preferences, opinions, and needs of those around them to avoid disharmony. They put others' needs first, suppressing their desires to maintain peace. This merging behavior leads to a loss of identity, making it difficult for Type 9s to develop a clear sense of self. Without asserting their own opinions or desires, they remain disconnected from their true identity, limiting their capacity for deeper self-awareness and growth. Over time, they can feel invisible or undervalued because they have lost touch with their motivations.

3. **Complacency and Numbing Out**

Type 9s can fall into patterns of inertia or complacency, avoiding activities or decisions that disrupt their comfort zone. They may distract themselves with routine, passive activities (e.g., watching TV or overeating) to avoid engaging with challenging or uncomfortable emotions. This numbing behavior dulls emotional consciousness. By avoiding stimulating or difficult experiences, Type 9s miss opportunities to engage with their emotions or reflect on their inner world. This avoidance can lead to emotional stagnation, where they no longer seek personal growth or self-awareness.

4. **Repression of Anger**

Type 9s repress anger because they associate it with conflict, which they fear. However, repressed anger can build up over time, leading to passive-aggressive behaviors or emotional withdrawal. Repressing anger disconnects Type 9s from their natural emotional responses. Consciousness requires the ability to experience and process all emotions, including anger. By suppressing this emotion, Type 9s lose access to critical internal signals that could inform them about boundaries, desires, or areas of dissatisfaction.

Type 9: Plasticity Challenges

Neuroplasticity refers to the brain's ability to reorganize itself by forming new neural connections in response to experiences, behaviors, and learning. Type 9s' preference for routine, avoidance of conflict, and resistance to discomfort can limit their brain's capacity to adapt and form new patterns.

Key Challenges:

1. **Inertia and Resistance to Change**

Type 9s tend to avoid taking initiative or making decisions, preferring to stay within their comfort zone. This resistance to change keeps them locked into familiar routines and passive behaviors. Neuroplasticity thrives on novelty and challenge. By avoiding new experiences or changes in their routine, Type 9s limit the brain's ability to form new neural connections. This inertia reinforces existing patterns, making it harder to shift out of passive behaviors or develop new ways of thinking.

2. **Lack of Stimulation**

Type 9s often gravitate toward passive activities that don't require much mental or emotional engagement. These can include mindlessly watching TV, zoning out, or other forms of disengagement. The brain requires stimulation to build new neural pathways. When Type 9s consistently engage in low-stimulation activities, they deprive their brain of the challenges needed to strengthen cognitive flexibility. Without mental stimulation, neuroplasticity diminishes, making it difficult for them to adapt to new situations or learn from experiences.

3. **Avoidance of Emotional Learning**

Type 9s may avoid uncomfortable emotions like anger, fear, or sadness, focusing instead on maintaining emotional numbness or contentment. They tend to suppress feelings that might disrupt their peace. Emotional learning is critical to neuroplasticity, allowing the brain to rewire itself based on emotional experiences. By avoiding emotional engagement, Type 9s miss the opportunity to reshape their emotional responses and develop emotional resilience. This limits their ability to create new, healthier emotional habits and patterns.

4. **Difficulty with Decision-Making**

Type 9s often struggle with making decisions because they fear that choosing one option may lead to conflict or disruption of harmony. As a result, they may procrastinate or defer decisions to others. Decision-making requires cognitive engagement and strengthens executive functioning in the brain. Avoiding decisions reduces the brain's ability to practice critical thinking, problem-solving, and weighing multiple perspectives, all essential for building new neural connections. Over time, this avoidance weakens neuroplasticity and reinforces passivity.

Type 9: Summary of Challenges

Challenge	Impact on Consciousness	Impact on Plasticity
Overactive Inner Critic	Prevents self-compassion and mindfulness; trapped in self-judgment.	Reinforces rigid thinking patterns and fear of imperfection.
Rigid Moral Compass	Limits openness to new perspectives; stuck in "right/wrong" thinking.	Inhibits cognitive flexibility; prevents exploration of new ideas.
Repression of Anger	Suppresses emotional awareness and processing.	Blocks emotional adaptability; creates unconscious emotional patterns.
Perfectionism	Creates tension and blocks acceptance of the present moment.	Hinders willingness to take risks or make mistakes.
Fear of Making Mistakes	This leads to avoidance of situations that challenge self-concept.	It limits the brain's ability to learn and adapt to errors.
Need for Control	Resists surrendering to the present, reducing flow and relaxation.	Restricts the brain's ability to adapt to change and uncertainty.

Type 9: PRIMAL Rituals Recommended

Grounding practices awaken Type 9s' energy and connect them to their physical self. These rituals soften conflict avoidance and encourage vulnerability by fostering engagement and emotional awareness.

Open up to Your Body

⇒ **Why:** Type 9s often struggle with inertia and

procrastination. Physical movement can re-energize them, helping to overcome feelings of stagnation or avoidance. Exercise also boosts mood, making it easier to engage with the world.

⇒ **Frequency:** Aim for daily movement, at least 30 minutes of moderate exercise (e.g., walking, yoga, swimming) to build energy and stay engaged.

Open up to Your Breath

⇒ **Why:** Breathwork helps Type 9s stay present and avoid slipping into numbing or disengagement. Mindful breathing grounds them in their body, increasing awareness of emotions they might otherwise avoid.

⇒ **Frequency:** Practicing 5-10 minutes of deep, mindful breathing exercises each morning or whenever stress or avoidance tendencies arise.

Open up to Nature

⇒ **Why:** Nature provides a peaceful, non-confrontational space for Type 9s to reset. Being outdoors can reduce stress, enhance mental clarity, and help them reconnect to their sense of calm and purpose.

⇒ **Frequency:** Spend at least 1-2 hours in nature each week. More extended periods, such as weekend hikes, are also beneficial for deepening their sense of serenity.

Type 9: RATIONAL Rituals Recommended

RATIONAL rituals guide Type 9s to assert themselves and prioritize their needs. By challenging complacency and embracing self-reflection, these practices reduce their tendency to merge with others and cultivate individuality.

Open up to Learning

⇒ **Why:** Type 9s can sometimes disengage mentally or feel

disconnected from their desires. Learning something new sparks curiosity, keeps their mind sharp, and helps them re-engage with the world meaningfully.

⇒ **Frequency:** Dedicate at least 30 minutes 2-3 times a week to learning a new subject or skill (e.g., reading, online courses, workshops).

Open up to Board Games

⇒ **Why:** These activities help keep Type 9s mentally alert and focused. Games or puzzles requiring problem-solving encourage active engagement and counter their tendency toward mental passivity or disengagement.

⇒ **Frequency:** Engage in puzzles or games for at least 1-2 hours per week, solo or with friends, to add a social element.

Open up to Cooking or Gardening

⇒ **Why:** These hands-on, creative activities help Type 9s overcome inertia and provide a tangible sense of accomplishment. These activities are also peaceful and allow for quiet contemplation, which aligns with their desire for calm.

⇒ **Frequency:** Aim for 2-3 weekly cooking or gardening sessions, each lasting at least an hour.

Type 9: SPIRITUAL Rituals Recommended

SPIRITUAL rituals nurture a more profound sense of purpose and connection for Type 9s. By engaging in contemplative practices, they can align with their inner truth and embrace the courage to engage fully in life.

Open up to Meditation

⇒ **Why:** Meditation helps Type 9s tap into their natural calm and serenity, pushing them to stay present rather than slipping into passive avoidance. It strengthens their awareness of their inner world without getting overwhelmed by it.

⇒ **Frequency:** A daily meditation practice of 10-15 minutes is ideal to build consistency and deepen awareness over time.

Open up to Mantras

⇒ **Why:** Type 9s can benefit from mantras or affirmations that reaffirm their worth and encourage them to act without sacrificing inner peace. Mantras can guide them through feelings of self-doubt or conflict avoidance.

⇒ **Frequency:** Practice using mantras or affirmations daily, especially at the start or end of the day. A few minutes of repeating affirmations like "I am capable of growth and change" can set a powerful tone

Type 9: Compatibility With Other Types

Type 1 (The Reformer)

•**What Works**: Type 9's easygoing nature can balance Type 1's perfectionism, offering patience and calm to their drive for improvement. Type 1 appreciates Type 9's non-confrontational attitude.
•**Potential Friction**: Type 1's criticism or rigid standards can frustrate Type 9, who avoids conflict. Type 9's passivity can irritate Type 1, who seeks action and precision.

Type 2 (The Helper)

•**What Works**: Both are people-oriented, and Type 2's giving nature complements Type 9's receptive and accommodating attitude. Together, they create a supportive, harmonious environment.
•**Potential Friction**: Type 2's emotional needs and desire for recognition may clash with Type 9's avoidance of deeper emotional engagement, leading to frustration for Type 2.

Type 3 (The Achiever)

•**What Works**: Type 9's calm presence provides stability and support for the ambitious Type 3. Type 9 admires Type 3's drive and success, while Type 3 appreciates Type 9's non-competitive nature.
•**Potential Friction**: Type 3's focus on achievement and image may overwhelm Type 9, who seeks peace and avoids high-pressure environments. Type 9 may feel overlooked or neglected by the success-driven Type 3.

Type 4 (The Individualist)

•**What Works**: Type 9's steady, non-judgmental nature can provide grounding for Type 4's emotional intensity. Type 4 appreciates Type 9's ability to listen and provide a calm space for self-expression.
•**Potential Friction**: Type 4's moodiness and need for deep emotional connection may frustrate Type 9, who avoids intense feelings. Type 9's tendency to suppress emotions may lead to emotional disconnect.

Type 5 (The Investigator)

•**What Works**: Both value peace and autonomy. Type 9's calm presence allows Type 5 the space they need, while Type 5's independence appeals to Type 9's desire to avoid confrontation and maintain harmony.
•**Potential Friction**: Type 9's avoidance of conflict may frustrate Type 5, who values intellectual engagement and clarity. Type 5's detachment can cause Type 9 to feel emotionally distant or unsupported.

Type 6 (The Loyalist)

•**What Works**: Both types value security and stability. Type 9's calm demeanor can ease Type 6's anxiety, while Type 6's loyalty provides the stability and reassurance that Type 9 craves.
•**Potential Friction**: Type 6's tendency to question and worry may

frustrate Type 9, who avoids overthinking and prefers peace. Type 9's indecisiveness or avoidance of conflict can increase Type 6's anxiety.

Type 7 (The Enthusiast)

•**What Works**: Both enjoy avoiding conflict and seeking pleasurable experiences. Type 9 appreciates Type 7's enthusiasm and ability to bring fun and excitement into their life, while Type 7 enjoys Type 9's calming energy.
•**Potential Friction**: Type 7's need for constant stimulation and change can overwhelm Type 9, who prefers stability and routine. Type 9's tendency to procrastinate or disengage may frustrate the fast-paced Type 7.

Type 8 (The Challenger)

•**What Works**: Type 9's calm, peaceful nature can soften Type 8's intensity. Type 8's strong leadership can give Type 9 a sense of protection and direction, while Type 9 provides patience and stability.
•**Potential Friction**: Type 8's aggressive style and desire for control may overwhelm Type 9, who avoids conflict at all costs. Type 9's passivity and resistance to action can frustrate Type 8's assertiveness.

Type 9 (The Peacemaker)

•**What Works**: Two Type 9s create a harmonious, low-conflict relationship in which both value peace and tranquility and share a mutual understanding of the need for calm and comfort.
•**Potential Friction**: Both may avoid making decisions or addressing conflicts, leading to stagnation or unresolved issues. The relationship can become too passive, neither pushing for growth nor change.

In summary, Type 9 pairs well with types that value harmony, stability, and support but may experience friction with emotionally intense types, overly critical, or confrontational. To thrive, Type 9s need partners who respect their need for peace but encourage them to engage more actively.

Type 9: Compatibility Table

Pair	Compatibility with Type 9	Strengths	Challenges	Tips for Improving Flow
9 and 1	Shared desire for harmony and doing what's right; Type 1's structure balances Type 9's adaptability.	Complementary balance between calmness (9) and discipline (1).	Type 1 may see Type 9 as too complacent; Type 9 may feel pressured by Type 1's high standards.	Type 9: Practice assertiveness to avoid being overly accommodating. Type 1: Build flexibility by relaxing perfectionism occasionally.
9 and 2	Warm, supportive relationship focused on care and harmony.	Both prioritize relationships and harmony.	Type 9 may withdraw when overwhelmed by Type 2's emotional intensity.	Type 9: Use plasticity to handle emotional engagement without retreating. Type 2: Practice self-care to avoid over-giving.
9 and 3	Type 9's calm steadiness complements Type 3's drive for achievement.	Creates a balance between ambition (3) and harmony (9).	Type 9 may feel left behind by Type 3's fast pace; Type 3 may view Type 9 as unmotivated.	Type 9: Develop goal-setting habits to match Type 3's energy. Type 3: Pause to appreciate Type 9's grounded perspective.

Pair	Compatibility with Type 9	Strengths	Challenges	Tips for Improving Flow
9 and 4	Emotional connection fosters deep understanding; Type 9 balances Type 4's emotional intensity.	Type 9's calmness can ground Type 4's emotional highs and lows.	Type 4 may feel unseen if Type 9 avoids conflict; Type 9 may find Type 4 overly dramatic.	Type 9: Expand consciousnes s by engaging more with Type 4's feelings. Type 4: Develop resilience to handle Type 9's calm passivity.
9 and 5	Quiet, harmonious relationship based on mutual respect for space and autonomy.	Both value calm and introspection, creating a peaceful connection.	Type 5 may find Type 9 unengaged intellectually; Type 9 may see Type 5 as emotionally distant.	Type 9: Build plasticity by engaging in intellectual discussions. Type 5: Practice openness to emotional sharing with Type 9.
9 and 6	Shared focus on security and relationships creates a strong foundation of trust.	Both value loyalty and reliability, fostering a sense of safety.	Type 6's anxiety may overwhelm Type 9; Type 9's indecision may frustrate Type 6.	Type 9: Use grounding habits to reassure Type 6. Type 6: Strengthen plasticity by encouraging Type 9's decision-making.

Pair	Compatibility with Type 9	Strengths	Challenges	Tips for Improving Flow
9 and 7	Fun, high-energy dynamic where Type 7 brings excitement and Type 9 provides grounding.	Type 9 tempers Type 7's impulsiveness, while Type 7 energizes Type 9.	Type 7 may feel bored by Type 9's slower pace; Type 9 may feel overwhelmed by Type 7's energy.	Type 9: Build plasticity by exploring Type 7's adventurous ideas. Type 7: Cultivate mindfulness to align with Type 9's calm energy.
9 and 8	Strong balance where Type 9's calmness softens Type 8's intensity.	Type 9 brings emotional balance to Type 8's bold nature; Type 8 inspires Type 9 to act.	Type 8 may overpower Type 9 in decision-making; Type 9 may withdraw from Type 8's assertiveness.	Type 9: Expand consciousnes s by asserting needs. Type 8: Practice vulnerability to connect emotionally with Type 9.
9 and 9	Extremely harmonious but may lack action or direction due to mutual avoidance of conflict.	Creates a serene, conflict-free environment with deep mutual understanding.	May avoid addressing issues, leading to stagnation or indecision.	Both: Develop plasticity by setting shared goals and practicing decision-making together.

Type 1: The Reformer (Analytical, idealistic, ethical, disciplined)

Principled and disciplined, Reformers are idealists driven by a strong desire to improve themselves and the world around them. They value integrity, order, and a sense of purpose. Each type develops unique coping mechanisms—patterns of thought, emotion, and behavior formed to address their core fears and navigate life's challenges. For Reformers, this often involves striving for perfection and rigidly adhering to standards to maintain control and avoid criticism. While these habits can fuel excellence, they may lead to burnout or self-criticism.

In the following sections, we explore how Type 1 typically operates in key areas of life—at work, home, and in relationships—revealing their characteristic strengths, vulnerabilities, and tendencies. We also examine this type's consciousness and plasticity challenges, identifying opportunities to enhance self-awareness, adaptability, and emotional intelligence. To support these efforts, we provide tailored *OPEN rituals*—PRIMAL, RATIONAL, and SPIRITUAL practices that nurture openness, balance, and personal growth.

Finally, the section concludes by unpacking compatibility dynamics, offering insights into how Type 1 relates to all other Enneagram types across various professional, familial, or romantic relationships.

The Gifts of Being an Enneagram 1

1. **Integrity and Responsibility:** One of the greatest strengths of Type 1s is their unwavering integrity. They hold themselves to high ethical standards and consistently strive to do what's right, which makes them reliable, trustworthy, and grounded.

2. **Commitment to Improvement:** Type 1s desire to improve the world (and themselves). Their pursuit of personal growth is inspiring, and they always work to improve their skills, knowledge, and relationships.

3. **Sense of Justice and Fairness**: Type 1s have a deeply rooted sense of justice and a solid moral compass. They are advocates

for fairness, equality, and doing things the right way, which makes them champions for positive change.

4. **Discipline and Work Ethic**: Their ability to stay focused and committed to their tasks is admirable. They are disciplined, hard-working, and highly organized, allowing them to excel in personal and professional areas.

Type 1 Coping Mechanisms

These mechanisms helped them survive and thrive in challenging environments, but they also created certain patterns that can present challenges in adulthood.
As children, Type 1s often internalized the belief that they needed to be "good" or "right" to be loved or accepted. This led to the development of several coping mechanisms, including:

1. Internalizing Criticism

- **Coping Mechanism:** As children, Type 1s often internalize high expectations from themselves or authority figures. They learned to self-criticize and strive for perfection to avoid disapproval or punishment. This need to be "good" and "right" helps them feel secure in a world they perceive as judgmental or chaotic.
- **Adult Challenge:** This internalized criticism becomes a harsh inner voice that constantly judges their actions and decisions. Type 1s may struggle with self-compassion as adults, leading to stress and burnout from their relentless pursuit of perfection.

2. Over-Control of Impulses

- **Coping Mechanism:** Type 1 children often suppress their natural impulses, believing spontaneity or imperfection could lead to mistakes or criticism. This self-control creates a sense of safety and ensures they will be considered responsible and disciplined.

Adult Challenge: This tendency to over-control can result in rigidity and difficulty enjoying simple pleasures. Type 1s may find it hard to relax or embrace spontaneity as adults, often feeling guilty for indulging in "unproductive" activities.

3. Hyper-Focus on Rules and Structure

- **Coping Mechanism:** Type 1 children coped by adhering strictly to rules or creating structures to maintain order in chaotic or unpredictable environments. They often became "mini-authorities", attempting to "fix" situations they perceived as flawed.
- **Adult Challenge:** As adults, this hyper-focus on rules can make Type 1s inflexible or overly critical of themselves and others. They may struggle in environments requiring ambiguity or collaboration, often feeling frustrated when things don't align with their ideal standards.

4. Suppressing Anger

- **Coping Mechanism:** Type 1 children often suppressed anger, believing it was "bad" or "unacceptable" to express negative emotions. Instead, they channeled this energy into striving for perfection or correcting perceived flaws.
- **Adult Challenge:** Suppressed anger can manifest as resentment or passive-aggressiveness in adulthood. Type 1s may find themselves bottling up frustration until it erupts unexpectedly, creating conflict in relationships and undermining their sense of balance.

The gifts of being an Enneagram Type 1 include a strong sense of integrity, discipline, and commitment to improving themselves and the world around them. They bring structure, order, and fairness into their lives, from work to relationships.

However, the coping mechanisms they developed in childhood—such as perfectionism, an overactive inner critic, and a need for control—can lead to inner tension, stress, and challenges in relationships. By recognizing these patterns and working to soften them, Type 1s can learn to balance their drive for excellence with self-compassion and flexibility, ultimately finding more peace and fulfillment.

Type 1 at Work

Strengths

⇒ **Perfection and High Standards**: Type 1s thrive in work environments that value high standards, attention to detail, and precision. They take their responsibilities seriously and have an inner drive to deliver excellent results.

⇒ **Integrity and Ethics**: Type 1s are deeply committed to doing things the right way, which makes them reliable employees, colleagues, or leaders. They often ensure that work is ethical and aligns with their sense of what's right.

⇒ **Order and Efficiency**: As natural organizers, Type 1s excel at creating structured systems, ensuring efficient workflows and project on-tracks.

Challenges

⇒ **Overwork and Rigidity**: The desire for perfection can make them overly critical of their work and others. They may feel pressure to over-deliver and do extra work to meet their standards.

⇒ **Difficulty Delegating**: Type 1s may struggle to trust others to perform tasks to their standards, leading them to micromanage or resist delegation.

⇒ **Judgment and Criticism**: Their internal criticism can become external, leading them to point out flaws or inefficiencies in colleagues, which can sometimes strain relationships at work.

Type 1 at Home

Strengths

⇒ **Creating Structure and Routine**: Type 1s often maintain a well-organized, structured home. They prefer order and cleanliness, creating a peaceful and efficient environment for themselves and their family.

⇒ **Responsibility and Dependability**: They take their role seriously, whether as parents, partners, or caretakers. They work hard to fulfill their duties and ensure that everyone's needs are met.

⇒ **Strong Moral Compass**: Type 1s often model strong ethics and values at home, encouraging honesty, fairness, and responsibility among family members.

Challenges

⇒ **Overly Controlling**: The need for order can lead Type 1s to enforce strict rules and routines, which can feel suffocating for others in the household.

⇒ **Unwillingness to Relax**: It can be difficult for a Type 1 to let go and relax when things are out of place or they feel something still needs to be done. They may struggle to balance work and play at home.

⇒ **High Expectations of Others**: Their inner critic often projects onto family members, making them overly critical of their partner or children, especially when things don't meet their expectations of "right."

Type 1 in Relationship

Strengths

⇒ **Commitment and Loyalty**: Type 1s are deeply loyal and committed to their relationships. They value their partners' integrity, honesty, and reliability and offer the same in return.

⇒ **Strong Sense of Justice:** They believe in fairness and will go to great lengths to ensure their relationships are based on respect and mutual support. They are often the moral compass in relationships, guiding their partners and loved ones toward ethical choices.

⇒ **Desire for Self-Improvement:** Type 1s always seek ways to grow and improve personally and in their relationships. They are willing to put in the effort to strengthen their relationships.

Challenges

⇒ **Criticism and Perfectionism:** The internal pressure to be perfect often leads Type 1s to judge themselves harshly, but this criticism can also spill over into their relationships. They may point out their partner's flaws or push them to meet high standards, creating tension.

⇒ **Difficulty Expressing Vulnerability:** Type 1s can struggle to show their softer side, feeling that vulnerability may be a sign of weakness. This can make it hard for them to fully open up emotionally.

⇒ **Desire for Control:** They may unconsciously try to control or "fix" their partner, believing they know the best way to handle things. This can create friction in relationships, especially if their partner feels judged or constrained.

Type 1 Consciousness Challenges

Consciousness refers to self-awareness, emotional intelligence, and the ability to observe one's thoughts and behaviors without being dominated by them. For Type 1, the core challenges related to consciousness and neuroplasticity stem from their deeply ingrained need for perfection, control, and the internalization of a critical inner voice. These challenges can limit their mental flexibility, growth, and self-awareness capacity. Let's explore these challenges more specifically:

1. **Overactive Inner Critic**

Type 1s often have a relentless inner critic that judges their thoughts, actions, and feelings. This inner voice can overshadow their conscious awareness, making it difficult to experience their emotions and behaviors without judgment. Instead of objectively observing their thoughts, they often evaluate them through a moral lens. The inner critic keeps them locked in a cycle of self-evaluation, preventing them from achieving a higher conscious awareness. Rather than being mindful and present, they are focused on correcting perceived flaws in themselves and others.

2. **Rigid Moral Compass**

Type 1s often see the world in black-and-white terms, leading them to believe there is a "right" way to do things. This moral rigidity can prevent them from being open to new perspectives or accepting things as they are because their focus is on how things *should* be. This rigidity inhibits the meta-awareness required for more profound levels of consciousness, where acceptance and non-judgment are crucial. It's hard for Type 1s to fully surrender to the present moment if they feel it doesn't meet their moral expectations.

3. **Repression of Anger**

Type 1s often suppress their anger because they view it as an "unacceptable" or "immoral" emotion. However, this repression can create internal frustration and block their ability to process and understand their feelings consciously. Repressed anger distorts their emotional consciousness. They may struggle to understand their emotional state or fully express it healthily. This can lead to unconscious emotional patterns emerging in unexpected or unhealthy behaviors.

Type 1 Plasticity Challenges

Plasticity refers to the brain's ability to adapt, learn, and reorganize itself in response to new experiences and information. Type 1s ' natural tendency toward perfectionism and rigid thinking patterns makes it more difficult to embrace change and adapt, thus limiting neuroplasticity.

Key Challenges:

1. **Perfectionism**

Type 1s tend to believe that there is a perfect way to do things, which can lead to rigidity in both thought and behavior. Their focus on making things "right" can prevent them from experimenting with new approaches or allowing themselves the freedom to make mistakes. Perfectionism directly inhibits plasticity because it limits the willingness to take risks, fail, or explore new possibilities. The brain thrives on new experiences, but Type 1s' perfectionism prevents them from embracing novelty due to fear of imperfection.

2. **Fear of Making Mistakes**

The fear of being wrong or making a mistake can paralyze Type 1s. This fear keeps them locked into repetitive behaviors and rigid patterns because they don't want to risk failure or criticism. Mistakes are essential for learning and rewiring the brain. When Type 1s avoid making mistakes, they deny themselves the opportunity for growth and brain adaptation. This rigidity reinforces fixed neural pathways rather than creating new, flexible ones.

3. **Difficulty with Flexibility**

Type 1s often have a fixed idea of how things should be, leading them to resist change or alternative approaches. They can become attached to specific rules, routines, and methods, struggling to adapt when circumstances shift. Plasticity requires cognitive flexibility, or the ability to adjust thinking patterns and behaviors in response to new challenges. Type 1s' difficulty with flexibility can stifle neuroplasticity because their brain isn't given the freedom to adapt to different ways of thinking or doing.

4. **Overemphasis on Control**

Type 1s often seek control over their environment and themselves to feel secure. They believe that if they maintain control, they can avoid chaos, disorder, or mistakes. Excessive control limits neuroplasticity because it restricts the brain's ability to adapt to change. True plasticity comes from allowing the brain to experiment and reorganize itself, but Type 1's need for control can prevent this process.

Type 1: Summary of Challenges

Challenge	Impact on Consciousness	Impact on Plasticity
Overactive Inner Critic	Prevents self-compassion and mindfulness; trapped in self-judgment.	Reinforces rigid thinking patterns and fear of imperfection.
Rigid Moral Compass	Limits openness to new perspectives; stuck in "right/wrong" thinking.	Inhibits cognitive flexibility; prevents exploration of new ideas.
Repression of Anger	Suppresses emotional awareness and processing.	Blocks emotional adaptability; creates unconscious emotional patterns.
Perfectionism	Creates tension and blocks acceptance of the present moment.	Hinders willingness to take risks or make mistakes.
Fear of Making Mistakes	Leads to avoidance of situations that challenge self-concept.	Limits brain's ability to learn and adapt from errors.
Need for Control	Resists surrendering to the present, reducing flow and relaxation.	Restricts brain's ability to adapt to change and uncertainty.

Type 1: PRIMAL Rituals Recommended

These rituals help Type 1s release their inner drive for perfection by connecting them to their physical body and the present moment. By softening their critical inner voice, these rituals encourage vulnerability and the acceptance of imperfection, grounding them in a sense of balance and emotional openness.

Open up to Your Breath

⇒ **Why:** Breathwork helps Type 1s release tension caused by perfectionism and regulate their nervous system. They can quiet their inner critic and reconnect with their body by focusing on their breath.

⇒ **Frequency:** Practice intentional breathwork for 5–10 minutes daily or whenever stress arises.

Open up to Nature

⇒ **Why:** Spending time in nature helps Type 1s escape their rigid mindset. Being in natural surroundings encourages calm, grounding, and an appreciation of imperfection.

⇒ **Frequency:** 30–60 minutes weekly, or incorporate shorter daily walks outdoors.

Open up to Your Body

⇒ **Why:** Gentle restorative movements, such as stretching or somatic exercises, help Type 1s release tension stored in their body from constant self-monitoring. This promotes relaxation and a sense of physical ease.

⇒ **Frequency:** 10–20 minutes daily or 2–3 times per week.

Type 1: RATIONAL Rituals Recommended

RATIONAL rituals for Type 1s focus on cultivating self-compassion and flexibility. Challenging rigid thinking and embracing vulnerability reduces the need for control and opens pathways to deeper emotional intelligence.

Open up to Evocative Music

⇒ **Why:** Listening to evocative music helps Type 1s balance their rational and emotional selves. It softens their inner critic by inspiring creativity and providing a sense of flow. Music encourages openness to imperfection and emotional expression.

⇒ **Frequency:** Listen to music daily, especially during moments of stress or while relaxing, for 10–15 minutes.

Open up to an Instrument

⇒ **Why:** Learning or playing an instrument encourages Type 1s to embrace practice and imperfection. It shifts their focus from outcomes to the joy of creative progress, fostering patience and self-compassion.

⇒ **Frequency:** Dedicate 20–30 minutes 2–3 times per week to playing or experimenting with an instrument.

Open up to Uplifting Movies

⇒ **Why:** Watching uplifting movies allows Type 1s to relax and connect with inspiring stories. It provides a mental break from self-criticism while expanding their perspective and appreciation for creativity.

⇒ **Frequency:** Watch an uplifting or inspiring movie once a week or explore shorter videos when time is limited.

Open up to Board Games

⇒ **Why:** Playing board games engages Type 1s in a structured yet playful activity, challenging their focus on perfection and inviting collaboration or strategic thinking. Games promote mental flexibility and joy.

⇒ **Frequency:** Play board games 1–2 times weekly for 30–45 minutes.

Open up to Art

⇒ **Why:** Creating or engaging with art helps Type 1s access their creativity while letting go of rigidity. Art fosters self-expression and encourages openness to imperfection and exploration.
⇒ **Frequency:** Spend 20–30 minutes 2–3 times weekly.

Type 1: SPIRITUAL Rituals Recommended

SPIRITUAL rituals for Type 1s emphasize surrendering their need for moral and personal correctness. Through rituals like meditation or prayer, they can align with a greater sense of purpose and let go of the ego-driven need for self-criticism.

Open up to Meditation

⇒ **Why:** Meditation helps Type 1s quiet their inner critic, embrace the present moment, and cultivate self-compassion. It allows them to observe thoughts without judgment, creating space for emotional balance and higher awareness.
⇒ **Frequency:** Practice meditation for 10–15 minutes daily or longer during moments of stress or reflection.

Open up to Prayers

⇒ **Why:** Prayer, whether in a traditional or personal sense, encourages Type 1s to connect with a higher power or sense of purpose. It provides a moment to release burdens and cultivate gratitude, fostering humility and inner peace.
⇒ **Frequency:** Spend 5–10 minutes daily -- morning or evening routine.

Open up to Your Dreams

⇒ **Why:** Reflecting on dreams allows Type 1s to explore their subconscious and gain insights into hidden fears or desires. It invites creativity and helps loosen the rigidity of their waking thoughts.

⇒ **Frequency:** Keep a dream journal by your bedside and reflect on dreams upon waking at least 3–4 times weekly.

Open up to Stargazing

⇒ **Why:** Stargazing offers Type 1s a sense of wonder and perspective, helping them release their focus on perfection and connect with the vastness of the universe. It fosters a sense of peace and awe.

⇒ **Frequency:** Spend 10–20 minutes stargazing weekly, especially in quiet, reflective moments.

Open up to Entheogens

⇒ **Why:** For Type 1s who feel ready, entheogens may provide profound experiences of self-transcendence, allowing them to connect with higher consciousness and release deep-seated self-critical patterns.

⇒ **Frequency:** Engage with entheogens only under proper guidance and in a safe, intentional setting as part of a transformative journey.

Open up to Mantras

⇒ **Why:** Reciting mantras helps Type 1s counter negative self-talk with affirmations of peace, love, and acceptance. It trains the mind to focus on uplifting truths rather than self-criticism.

⇒ **Frequency:** Recite 5–10 minutes daily or incorporate mantras into meditation or prayer practices.

Type 1: Compatibility With Other Types

Type 1 (The Reformer)

•**What Works**: Two Type 1s share similar values, striving for integrity, discipline, and improvement. They respect each other's dedication to doing what is right and can work effectively toward shared goals.

•**Potential Friction**: Both may become overly critical of each other, amplifying perfectionist tendencies. Their mutual need for correctness could lead to power struggles over whose approach is more "right. "

Type 2 (The Helper)

•**What Works**: Both are caring and value helping others. Type 1 appreciates Type 2's generosity and desire to do good, while Type 2 admires Type 1's dedication to integrity and ethics.
•**Potential Friction**: Type 1's critical nature can clash with Type 2's emotional need for appreciation, making Type 2 feel undervalued. Type 1 might view Type 2's emotional approach as distracting from their sense of duty.

Type 3 (The Achiever)

•**What Works**: Both are driven and focused on achieving high standards. Type 1 admires Type 3's ambition, and Type 3 values Type 1's moral dedication to doing things right. Together, they can be a powerful, goal-oriented duo.
•**Potential Friction**: Type 1 may view Type 3's focus on success and image as superficial, while Type 3 may find Type 1 too rigid and overly critical of their approach to achieving results.

Type 4 (The Individualist)

•**What Works**: Type 4's creativity can inspire Type 1, bringing emotional depth to their structured approach. Type 1 helps Type 4 bring discipline to their artistic expression.
•**Potential Friction**: Type 4's emotional intensity and self-focus can frustrate Type 1's desire for order and consistency. Type 1's tendency to critique may make Type 4 feel misunderstood or constrained.

Type 5 (The Investigator)

•**What Works**: Both value knowledge, clarity, and precision. Type 1

appreciates Type 5's analytical thinking, while Type 5 respects Type 1's disciplined and principled approach to life.
•**Friction**: Type 5's tendency to detach emotionally may be perceived by Type 1 as a lack of engagement or responsibility. Type 1's insistence on action and improvement may overwhelm Type 5's preference for observation and reflection.

Type 6 (The Loyalist)

•**What Works**: Both are responsible and conscientious. Type 1 values Type 6's loyalty and dependability, while Type 6 respects Type 1's moral clarity and structured approach to problem-solving.
•**Potential Friction**: Type 6's anxiety and tendency to second-guess can frustrate Type 1's desire for decisive action. Type 1's critical nature might exacerbate Type 6's worries, making them feel judged or pressured.

Type 7 (The Enthusiast)

•**What Works**: Type 7 brings joy and spontaneity to Type 1's structured life, helping them loosen up and enjoy the moment. Type 1 provides the discipline and focus that Type 7 often lacks.
•**Potential Friction**: Type 1's strictness and need for order may clash with Type 7's desire for freedom and avoidance of responsibility. Type 7 might feel stifled by Type 1's focus on rules and improvement.

Type 8 (The Challenger)

•**What Works**: Both are assertive and action-oriented, with a shared desire for justice and fairness. Type 1 respects Type 8's strength and decisiveness, while Type 8 values Type 1's moral convictions.
•**Potential Friction**: Type 8's confrontational and aggressive approach can overwhelm Type 1's desire for proper conduct and order. Type 1 may find Type 8 too forceful, while Type 8 may feel that Type 1 is overly critical or rigid.

Type 9 (The Peacemaker)

•**What Works**: Type 9's calm, easygoing nature complements Type 1's drive for order, creating a peaceful environment where both can thrive. Type 1 offers structure, while Type 9 provides harmony.
•**Potential Friction**: Type 9's tendency to avoid conflict and decision-making can frustrate Type 1, who seeks action and resolution. Type 1's criticism may cause Type 9 to withdraw or become passive-aggressive.

Type 1: Compatibility Table

Pair	Compatibility with Type 1	Strengths	Challenges	Tips for Improving Flow
1 and 2	Shared desire to help others fosters a strong partnership.	Type 1's structure complements Type 2's emotional warmth.	Type 2 may feel criticized by Type 1's high standards; Type 1 may feel burdened by Type 2's need for appreciation.	Type 1: Raise consciousness by practicing patience with Type 2's emotions. Type 2: Build plasticity by focusing on self-care.
1 and 3	Ambitious, goal-oriented duo that excels in accomplishing tasks.	Type 1's perfectionism and Type 3's drive create effective teamwork.	Type 1 may find Type 3 too focused on image; Type 3 may feel restricted by Type 1's rigid standards.	Type 1: Practice flexibility by valuing Type 3's pragmatic approach. Type 3: Expand consciousness by embracing Type 1's ethical vision.

Pair	Compatibility with Type 1	Strengths	Challenges	Tips for Improving Flow
1 and 4	Emotional depth from Type 4 balances Type 1's rationality, fostering creativity and integrity.	Type 4's creativity enhances Type 1's structured thinking.	Type 4 may feel judged by Type 1's critical nature; Type 1 may struggle with Type 4's moodiness.	Type 1: Develop plasticity by embracing Type 4's emotional flow. Type 4: Expand consciousness by focusing on shared values with Type 1.
1 and 5	Intellectually stimulating pair, where Type 5's curiosity complements Type 1's precision.	Both value knowledge and careful decision-making.	Type 5 may find Type 1 too rigid; Type 1 may see Type 5 as overly detached.	Type 1: Build plasticity by loosening rigid thinking. Type 5: Expand consciousness by engaging emotionally with Type 1's ideals.
1 and 6	Shared focus on integrity and responsibility creates a reliable partnership.	Both are loyal and committed to doing what's right.	Type 6's anxiety may clash with Type 1's need for order; Type 1 may see Type 6 as overly cautious.	Type 1: Raise consciousness by offering reassurance to Type 6. Type 6: Develop plasticity by trusting Type 1's structured approach.

Pair	Compatibility with Type 1	Strengths	Challenges	Tips for Improving Flow
1 and 7	Type 7's optimism lightens Type 1's seriousness, fostering balance and creativity.	Type 7 inspires Type 1 to embrace spontaneity, while Type 1 grounds Type 7's adventurous nature.	Type 1 may find Type 7 irresponsible; Type 7 may see Type 1 as overly critical or restrictive.	Type 1: Build plasticity by embracing spontaneity. Type 7: Expand consciousness by valuing Type 1's sense of purpose.
1 and 8	A powerful, action-oriented duo focused on justice and results.	Type 8's decisiveness aligns with Type 1's principled nature.	Type 1 may feel bulldozed by Type 8's intensity; Type 8 may find Type 1 overly critical or slow to act.	Type 1: Expand consciousness by appreciating Type 8's boldness. Type 8: Build plasticity by valuing Type 1's ethical perspective.
1 and 9	Type 1's structure balances Type 9's desire for harmony, fostering stability and mutual respect.	Type 9's calmness tempers Type 1's intensity, creating balance in decision-making.	Type 1 may feel frustrated by Type 9's indecision; Type 9 may feel pressured by Type 1's high expectations.	Type 1: Raise consciousness by practicing patience with Type 9's pace. Type 9: Build plasticity by engaging actively in decision-making.

Pair	Compatibility with Type 1	Strengths	Challenges	Tips for Improving Flow
1 and 1	Shared values and commitment to ethics make for a highly principled partnership.	Both strive for improvement and share a deep sense of integrity.	Risk of excessive criticism and inflexibility in the relationship.	Both develop plasticity by practicing self-compassion and loosening rigid expectations of each other.

THE HEART TRIAD

HEART Types (2,3,4): Fear of Shame

These types are concerned with identity, recognition, shame, and worthiness. Their core fear is not being seen and loved. Each type manages this fear differently: Type 2 by gaining love through helping, Type 3 through achievement, and Type 4 through individuality.

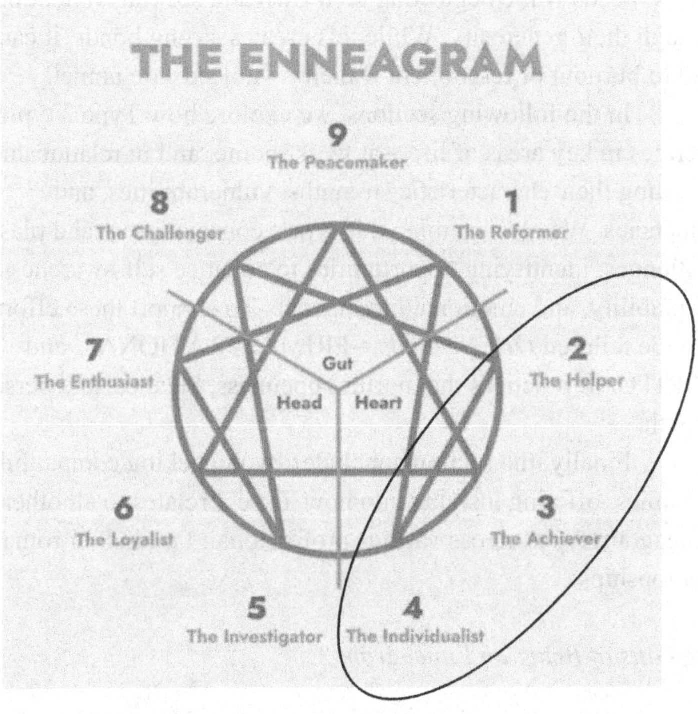

- **Type 2 (The Helper):** Their core fear is being unloved or unwanted by themselves. They seek to mitigate this fear by becoming indispensable to others.

- **Type 3 (The Achiever):** They fear being worthless or failing to be seen as flourishing and valuable. This fear drives their pursuit of achievement and recognition.

- **Type 4 (The Individualist):** Their core fear is of having no identity or personal significance. They respond to this fear by cultivating a unique identity.

Type 2: The Helper (Compassionate, relational, giving, attentive)

Empathetic and giving, Helpers are driven by a deep need to connect with others and feel valued through acts of service and care. They thrive on building relationships and earning appreciation. Each type develops unique coping mechanisms—patterns of thought, emotion, and behavior formed to address their core fears and navigate life's challenges. For Helpers, this often means focusing on others' needs while neglecting their own and seeking validation through their generosity. While this creates strong bonds, it can also lead to burnout or resentment if their own needs are unmet.

In the following sections, we explore how Type 2 typically operates in key areas of life—at work, home, and in relationships—revealing their characteristic strengths, vulnerabilities, and tendencies. We also examine this type's consciousness and plasticity challenges, identifying opportunities to enhance self-awareness, adaptability, and emotional intelligence. To support these efforts, we provide tailored *OPEN rituals*—PRIMAL, RATIONAL, and SPIRITUAL practices that nurture openness, balance, and personal growth.

Finally, the section concludes by unpacking compatibility dynamics, offering insights into how Type 2 relates to all other Enneagram types across various professional, familial, or romantic relationships.

The Gifts of Being an Enneagram 2

1. **Empathy and Compassion:** Type 2s have an extraordinary ability to deeply understand and empathize with others. They can intuitively sense when someone is in need, emotionally or physically, and are often the first to offer comfort and support. Their compassion allows them to connect with people profoundly, offering genuine care and kindness.

2. **Emotional Intelligence:** Type 2s are incredibly in tune with their own emotions and those of others. They can read between the lines, understanding the unspoken feelings of those around them. This emotional intelligence makes them great friends,

partners, and colleagues, as they can respond to people in ways that meet their emotional needs.

3. **Generosity and Selflessness:** A defining trait of Type 2s is their genuine desire to help and support others. They derive great joy and fulfillment from giving, often putting others' needs before theirs. Whether offering a helping hand, providing emotional support, or simply being present for someone in need, Type 2s are incredibly generous with their time, love, and resources.

4. **Dedication to Service:** Type 2s are highly motivated by the desire to make a positive difference in the lives of others. They find purpose in being of service and are often driven by a deep need to help, heal, or nurture those around them. This can manifest in personal and professional settings, where they excel in caregiving, teaching, or counseling roles.

5. **Positive Influence on Relationships:** Type 2s are the glue that holds people together in personal and professional relationships. They foster open communication, emotional warmth, and trust, creating an environment where relationships thrive. Their natural inclination to care for others helps resolve conflicts and strengthens bonds.

6. **Optimism and Uplifting Energy:** Despite their focus on others' needs, Type 2s often exude an uplifting and positive energy that can brighten the lives of those around them. They help others see the good in difficult situations and are usually the ones to encourage and inspire hope during tough times.

7. **Creating a Sense of Belonging:** Type 2s have a gift for making others feel important and appreciated. They are excellent at ensuring that people feel included and loved, which helps to create a strong sense of belonging within a group. Whether in their family, friendships, or workplace, they work tirelessly to foster environments where everyone feels valued.

Type 2 Coping Mechanisms

Type 2s develop coping mechanisms to navigate their deep desire for connection and need to feel loved and valued. These mechanisms often stem from the belief that their worth is tied to how much they give or are needed.

1. Over-Giving to Others

- **Coping Mechanism:** As children, Type 2s may have learned to gain approval and love by being overly helpful and attuned to others' needs. They became "caretakers" to feel valued and secure, often putting aside their needs.
- **Adult Challenge:** Over-giving can lead to exhaustion and resentment in adulthood when others do not reciprocate or acknowledge their efforts. Type 2s may struggle to set boundaries and prioritize self-care, fearing rejection if they are not constantly giving.

2. Suppressing Personal Needs

- **Coping Mechanism:** Type 2s often ignored their own needs as children, believing they were less important than the needs of others. By focusing on making others happy, they felt safer and more accepted.
- **Adult Challenge:** Suppressing their needs can lead to burnout and feeling unappreciated. In adulthood, Type 2s may struggle with articulating their desires and asking for help, fearing appearing selfish or unworthy of love.

3. Seeking Validation Through Helping

- **Coping Mechanism:** Type 2s often sought validation by being indispensable to others. They learned that being helpful or kind was a way to secure affection and avoid rejection.
- **Adult Challenge:** This reliance on external validation can cause Type 2s to base their self-worth on others' approval. If their help is unnecessary, they may feel unworthy or

unloved, leading to emptiness or insecurity.

4. Emotional Manipulation to Gain Connection

- **Coping Mechanism:** Growing up, Type 2s often felt neglected or unloved. As a result, they may have learned to use subtle emotional manipulation—such as guilt or praise—to maintain closeness with others.
- **Adult Challenge:** In adulthood, this behavior can create unhealthy dynamics in relationships. Type 2s may unintentionally foster codependency or feel hurt when their efforts to secure connection are not reciprocated.

5. Difficulty Expressing Vulnerability

- **Coping Mechanism:** To maintain their role as helpers or nurturers, Type 2s often avoid showing vulnerability, believing they must be strong for others.
- **Adult Challenge:** Suppressing vulnerability can lead to a lack of emotional intimacy in relationships. Type 2s may struggle to share their struggles or ask for support, perpetuating feelings of isolation and unfulfilled emotional needs.

The gifts of Type 2s—empathy, generosity, emotional intelligence, and the ability to build deep connections make them invaluable in every aspect of life. They bring love, support, and compassion wherever they go, nurturing the emotional and relational well-being of everyone they encounter. By fully embracing their gifts while learning to set healthy boundaries and care for themselves, Type 2s can experience true fulfillment and create lasting, meaningful change in their relationships and communities.

The challenges that arise from Type 2's coping mechanisms—such as overgiving, neglecting self-care, and seeking validation through service—can lead to emotional exhaustion, resentment, and a lack of fulfillment. By recognizing these tendencies and learning to set healthy boundaries, Type 2s can begin to balance their natural generosity with self-care, ultimately fostering healthier, more fulfilling relationships with them and others. The key for Type 2s is learning to love and care for themselves with the compassion they offer others.

Type 2 at Work

Strengths

⇒ **Support and care of others:** Type 2's are driven by a desire to be helpful and appreciated. You often go above and beyond to help colleagues, take on additional tasks, and create a warm, harmonious environment. Type 2s excel in roles directly assisting others, such as teaching, counseling, or healthcare. They are likely to be the person who remembers everyone's birthday, check in on colleagues' well-being, and help even before it's requested.

⇒ **Warm and collaborative:** Type 2s bring empathy and a strong sense of teamwork to the workplace. You excel in creating an inclusive and supportive environment where people feel cared for and valued.

Challenges

⇒ **Overextension:** Type 2s to be needed can lead to overextension, where you take on too much responsibility, risking burnout. Type 2s may also struggle with boundary-setting, finding it difficult to say no or ask for help themselves.

Type 2 at Home

Strengths

⇒ **Holds family together**: Type 2s are attentive to the needs of those around them, ensuring everyone is comfortable and cared for, whether cooking meals, organizing family gatherings, or offering emotional support.

⇒ **Creator of a warm environment:** They thrive on creating a loving, supportive home environment. Their selflessness makes them natural caregivers within the household.

Challenges

- ⇒ **Neglecting self-care**: focusing on others' needs can sometimes lead to neglecting your self-care
- ⇒ **Resentment:** Type 2s prioritize others' needs over their own, leading to resentment or exhaustion, especially if their efforts go unappreciated. They may also struggle to communicate their needs, expecting others to understand them intuitively.

Type 2 in Relationship

Strengths

- ⇒ **Deeply caring:** Type 2s express their love through acts of service, always looking for ways to support their loved ones and make their lives easier.
- ⇒ **Compassionate:** They bring loyalty, and deep emotional intelligence to relationships. They are attentive to their partner's emotional needs, offering constant support and affection.

Challenges

- ⇒ **Fear of being unloved**: Type 2s work hard to ensure their partner feels valued, cherished, and emotionally vulnerable in relationships. However, this intense focus on the other person's needs can sometimes backfire, leading to codependency or feeling unappreciated if their efforts aren't reciprocated.
- ⇒ **Overgiving:** Type 2's need to feel needed can lead to overgiving, where they may sacrifice their well-being or emotional needs to please their partner. Additionally, they might feel unappreciated if their efforts aren't acknowledged, which can create emotional tension.

Type 2 Consciousness Challenges

Type 2s, the compassionate and giving "Helpers," often focus so intently on others' needs that they neglect their inner awareness. Their strong desire for connection and approval can make recognizing and addressing their boundaries and emotions challenging. This outward focus may hinder their ability to cultivate self-awareness and achieve balance in their relationships and inner life.

1. **Avoidance of Personal Needs**

Type 2s tend to focus so much on meeting the needs of others that they often ignore or suppress their desires and emotions. This tendency challenges developing consciousness because self-awareness requires recognizing and addressing one's inner world. By constantly directing their attention outward, Type 2s may struggle to connect with their deeper emotional needs or even deny that they have any.

2. **Emotional Dependency on External Validation**

A significant challenge for Type 2s is relying on external validation to feel worthy or loved. They often gauge their value based on how much others need them. This dependency on external sources of love can block the growth of consciousness, as it keeps Type 2s focused on others' reactions rather than developing an internal sense of self-worth. Emotional dependency limits self-awareness and self-love. Type 2s may spend so much time seeking affirmation that they overlook opportunities for deeper emotional understanding and self-acceptance.

3. **Difficulty with Emotional Boundaries**

Type 2s often blur the lines between their emotions and those of others, complicating consciousness development. They may feel overwhelmed by others' feelings without clear emotional boundaries, making distinguishing their emotions from external influences difficult. This enmeshment hampers the ability to step back and objectively observe their inner emotional landscape. Lack of boundaries creates confusion in their emotional world, making it hard for Type 2s to see their true feelings and thoughts, which hinders personal consciousness growth.

4. **Reluctance to Face Negative Emotions**

Type 2s prefer to focus on positive interactions and may shy away from uncomfortable or negative emotions. This avoidance of discomfort can limit the growth of consciousness, which requires engaging with all emotional experiences, even the painful ones. To develop higher consciousness, it is essential to confront and process negative feelings rather than ignore or downplay them. By avoiding negative emotions, Type 2s prevent themselves from reaching a deeper level of emotional insight and awareness, limiting their overall growth in consciousness.

5. **Subconscious Manipulation for Love**

A more profound unconscious challenge for Type 2s is their tendency to use giving or caretaking to receive love and validation, often without realizing it. This subconscious manipulation can inhibit self-awareness, as Type 2s may not be fully conscious of how much their actions are driven by their own needs for affection and approval. This lack of self-awareness around motives creates blind spots in Type 2s' consciousness, preventing them from fully understanding why they behave the way they do in relationships.

Type 2: Plasticity Challenges

Neuroplasticity refers to the brain's ability to reorganize itself by forming new neural connections throughout life. It allows people to change habits, beliefs, and emotional responses through new experiences and learning.

Key Challenges:

1. **Over-focus on External Validation**

Type 2s' fixation on pleasing others limits their ability to develop new, healthy neural patterns, which is essential for plasticity. When they continually focus on external approval, they miss opportunities to practice self-directed growth or self-soothing, reinforcing old neural pathways prioritizing others' needs over their own. This overdependence on external validation creates rigid, habitual patterns in the brain that inhibit flexibility and growth, preventing Type 2s from developing healthier, self-sustaining emotional habits.

2. **Resistance to Self-Care and Self-Prioritization**

Plasticity involves the brain's ability to adapt and form new neural connections through new experiences and habits. For Type 2s, however, their reluctance to prioritize self-care prevents them from creating habits that promote personal growth. Because they focus on others, they miss opportunities to develop practices that enhance their emotional and mental well-being. Avoiding self-care inhibits forming new, healthier neural patterns essential for emotional resilience and mental flexibility. This can trap Type 2s in the same emotional loops of over-giving and neglecting themselves.

3. **Fear of Abandoning Old Roles**

Type 2s often struggle with plasticity because they fear that if they stop their caregiving role, they will no longer be needed or loved. This fear keeps them locked into old patterns of behavior that may have been useful in childhood but are now limiting their personal and emotional development. Fear of losing the caregiver identity makes it difficult for Type 2s to embrace new behaviors and emotional responses, which are essential for neural growth and plasticity. This rigidity can prevent them from forming new, healthier ways of interacting in relationships.

4. **Difficulty with Emotional Detachment**

Plasticity thrives when we learn to detach and observe our patterns objectively and emotionally. For Type 2s, who are deeply enmeshed in others' emotions, this emotional detachment is complex. Their habitual merging with others' emotional states can prevent them from stepping back and developing new cognitive or emotional frameworks. Emotional entanglement makes it hard for Type 2s to build new mental patterns that would allow for healthier boundaries. Without emotional separation, they remain stuck in old emotional reactions that limit their brain's flexibility and adaptation.

5. **Over-Giving Prevents New Experiences**

Type 2s often dedicate so much of their time and energy to supporting others that they miss opportunities for new, enriching experiences that could boost their brain's plasticity. New experiences are critical for plasticity as they help the brain build new pathways and connections, but Type 2s may not take the time for self-enriching experiences. By limiting their opportunities for growth and novelty, Type 2s may stagnate in terms of neural development. Over-giving narrows their exposure to new experiences, which is essential for keeping the brain adaptable and resilient.

Type 2: Summary of Challenges

Challenge	Impact on Consciousness	Impact on Plasticity
Avoidance of Personal Needs	Prevents self-awareness and emotional insight, hindering personal growth.	Limits development of self-care habits, reinforcing old neural patterns of over-giving.
Emotional Dependency on Validation	Keeps focus on external approval, reducing self-worth and self-acceptance.	Reinforces habitual patterns of seeking external validation, inhibiting brain flexibility.
Difficulty with Emotional Boundaries	Creates confusion between personal emotions and others', blocking introspection.	Prevents formation of new, healthier boundaries, leading to emotional enmeshment.
Reluctance to Face Negative Emotions	Avoids deeper emotional engagement, limiting self-awareness.	Limits neural growth by avoiding the emotional discomfort necessary for change.
Subconscious Manipulation for Love	Creates blind spots in motivation and emotional needs, hindering consciousness.	Reinforces manipulative patterns, preventing new emotional responses from developing.
Resistance to Self-Prioritization	Blocks self-reflection and personal growth by focusing on others' needs.	Hinders the formation of self-care practices that are key to building brain plasticity.
Fear of Abandoning Old Roles	Keeps Type 2 stuck in old identities, preventing deeper self-awareness.	Limits adaptability by reinforcing outdated behavioral patterns.

Challenge	Impact on Consciousness	Impact on Plasticity
Over-Giving Prevents New Experiences	Focus on others limits opportunities for personal growth and awareness.	Reduces exposure to new experiences, limiting neural connections and adaptability.

Type 2: PRIMAL Rituals Recommended

Grounding rituals for Type 2s focus on reclaiming their sense of self. These rituals soften their compulsion to help others constantly and encourage emotional vulnerability by focusing on their own needs and physical well-being.

Open up to Your Body

⇒ **Why**: Encourages Type 2s to reconnect with their physical self, promoting self-care and emotional balance.
⇒ **Frequency**: Daily.

Open up to Your Breath

⇒ **Why**: Supports emotional regulation and relaxation, helping Type 2s manage stress and avoid burnout.
⇒ **Frequency**: Daily for 5-10 minutes.

Open up to Nature

⇒ **Why**: Immersion in nature reduces emotional fatigue, offers grounding, and fosters self-renewal.
⇒ **Frequency**: Weekly.

Open up to Your Sleep

⇒ **Why**: Ensures emotional resilience by providing the rest needed for self-regulation and personal energy.

⇒ **Frequency**: Nightly.

Open up to Your Food

⇒ **Why**: Shifts focus to personal nourishment, helping Type 2s care for their body with intentional eating.

⇒ **Frequency**: Each meal.

Type 2: RATIONAL Rituals Recommended

RATIONAL rituals encourage Type 2s to explore boundaries and embrace self-awareness. These practices challenge their need to earn love through service and invite them to prioritize self-care, fostering authentic emotional connections.

Open up to Learning

⇒ **Why**: Expanding knowledge fosters self-reflection, reduces over-dependence on others, and nurtures personal growth.

⇒ **Frequency**: Weekly.

Open up to Evocative Music

⇒ **Why**: Evocative music enables emotional expression and introspection, providing a creative outlet.

⇒ **Frequency**: Weekly.

Open up to Cooking

⇒ **Why**: A creative approach to cooking allows Type 2s to nourish themselves while enjoying the process.

⇒ **Frequency**: Several times a week.

Open up to Board Games

⇒ **Why**: Encourages lighthearted social connection without overextending emotionally.

⇒ **Frequency**: Weekly.

Open up to Traveling

⇒ **Why**: Exploring new places provides a sense of renewal and helps Type 2s embrace their independence.

⇒ **Frequency**: Periodically, as opportunities arise.

Type 2: SPIRITUAL Rituals Recommended

For Type 2s, SPIRITUAL rituals create a safe space to explore their innate worth without external validation. Rituals like silent reflection or chanting release the ego's need for approval and deepen their connection to their higher self.

Open up to Meditation

⇒ **Why**: Meditation fosters inner peace and helps Type 2s develop a stronger connection to themselves.

⇒ **Frequency**: Daily, 10-15 minutes.

Open up to Prayers

⇒ **Why**: Anchors Type 2s in gratitude and helps them align with a higher sense of purpose.

⇒ **Frequency**: As needed, especially during stressful times.

Open up to Stargazing

⇒ **Why**: Stargazing shifts focus from others' needs to a larger perspective, cultivating awe and wonder.

⇒ **Frequency**: Monthly or during clear nights.

Open up to Your Dreams

⇒ **Why**: Reflecting on dreams helps uncover subconscious desires and unresolved emotions.
⇒ **Frequency**: Nightly journaling upon waking.

Open up to Mantras

⇒ **Why**: Repeating affirmations builds inner confidence and reinforces emotional boundaries.
⇒ **Frequency**: Daily.

Type 2: Compatibility With Other Types

Enneagram Type 2, "The Helper," is warm, compassionate, and driven by the desire to support and nurture others. They excel at building emotional connections but can sometimes overextend themselves in relationships. Let's explore how Type 2 interacts with the other Enneagram types, including what works easily and where friction may arise.

Type 1 (The Reformer)

•**What Works**: Type 1's strong sense of responsibility and ethics pairs well with Type 2's nurturing nature. Type 2s appreciate the stability and structure Type 1 brings, while Type 1s value the warmth of Type 2s.
•**Potential Friction**: Type 1's perfectionism and critical tendencies may clash with Type 2's need for appreciation and validation. Type 1s may see Type 2 as overly emotional, while Type 2s could feel unappreciated or judged by Type 1's high standards.

Type 2 (The Helper)

•**What Works**: Two Type 2s create a warm, supportive, and nurturing relationship. Both partners will likely be highly attuned to each other's needs, creating a strong emotional bond.
•**Potential Friction**: Both may overextend themselves in meeting each other's needs, leading to exhaustion or burnout. There's also the possibility of competition for appreciation or validation.

Type 3 (The Achiever)

•**What Works**: Type 2's supportive and encouraging nature complements Type 3's drive for success. Type 2s provide the emotional backing and care that allows Type 3s to shine in their goals and achievements.
•**Potential Friction**: Type 3's focus on ambition and external success might leave Type 2 feeling neglected or underappreciated. Type 2s may also feel that Type 3 is too focused on image, while Type 3 could view Type 2 as overly dependent on others' appreciation.

Type 4 (The Individualist)

•**What Works**: Type 2's nurturing care can help Type 4 feel seen and understood, while Type 4's emotional depth brings richness to the relationship. Both value emotional intimacy, creating a strong connection.
•**Potential Friction**: Type 2's tendency to avoid their own emotional needs could clash with Type 4's intense focus on emotions. Type 4 may see Type 2 as avoiding their inner depths, while Type 2 may find Type 4's emotional intensity overwhelming.

Type 5 (The Investigator)

•**What Works**: Type 2's warmth and caring nature can draw Type 5 out of their intellectual world, creating a balance between emotional and intellectual connection. Type 5s appreciate the nurturing attention of Type 2s.
•**Potential Friction**: Type 5's need for independence and space may conflict with Type 2's desire for closeness and emotional connection. Type 2s may feel rejected or emotionally distant from Type 5, while Type 5s might feel overwhelmed by Type 2's needs.

Type 6 (The Loyalist)

•**What Works**: Type 2's supportive nature aligns well with Type 6's

need for security and loyalty. Both value emotional stability and loyalty in relationships, creating a strong foundation of trust.
•**Potential Friction**: Type 6's anxiety and need for reassurance can sometimes overwhelm Type 2, who may feel that their support is never enough. Conversely, Type 6 might feel smothered by Type 2's attempts to help and could pull away to create distance.

Type 7 (The Enthusiast)

•**What Works**: Type 2's nurturing energy complements Type 7's adventurous spirit, creating a dynamic relationship where Type 7 brings excitement and Type 2 offers emotional support.
•**Potential Friction**: Type 7's desire for freedom and avoidance of emotional depth may frustrate Type 2, who seeks emotional closeness and stability. Type 2 might feel neglected if Type 7 focuses too much on their fun and not enough on the relationship.

Type 8 (The Challenger)

•**What Works**: Type 8's confidence and assertiveness complement Type 2's nurturing tendencies. Type 2 feels protected and valued by Type 8's strength, while Type 8 benefits from Type 2's emotional care and support.
•**Potential Friction**: Type 8's direct and sometimes domineering nature could overpower Type 2, leaving them feeling unappreciated or pushed aside. Type 2 may feel hurt by Type 8's bluntness, while Type 8 may find Type 2's emotional needs overwhelming.

Type 9 (The Peacemaker)

•**What Works**: Type 2 and Type 9 can create a harmonious, peaceful relationship where both seek emotional connection and mutual support. Type 9's calm and accommodating nature helps Type 2 feel valued, while Type 2 provides the emotional nurturing that Type 9 appreciates.
•**Potential Friction**: Both types can avoid conflict, which might result in issues being swept under the rug rather than addressed. Type 2 may also overextend themselves trying to meet Type 9's needs,

while Type 9 may struggle to express their desires, leading to frustration.

Type 2s generally form strong, emotionally connected relationships, but their desire for appreciation and closeness can create friction with more independent or emotionally distant types. Their nurturing nature aligns well with types that value connection and loyalty, but they may need to establish better boundaries to avoid overextending themselves in relationships.

Type 2: Compatibility Table

Pair	Compatibility with Type 2	Strengths	Challenges	Tips for Improving Flow
2 and 1	Type 2's nurturing complements Type 1's focus on integrity and improvement.	Harmonious teamwork; 2's empathy balances 1's idealism.	2 may feel unappreciated; 1 may see 2 as overly emotional.	Cultivate mutual appreciation and flexibility in emotional expression.
2 and 2	Both share a deep desire to connect and care for others.	A shared love for giving creates emotional closeness.	Risk of overgiving; may compete for validation.	Set boundaries and prioritize self-care to avoid burnout.
2 and 3	Type 2 supports 3's ambition, while 3 appreciates 2's encouragement.	Inspiring partnership; 2 nurtures while 3 motivates.	3's focus on success may overlook 2's emotional needs.	Balance achievement with authentic emotional connection.

Pair	Compatibility with Type 2	Strengths	Challenges	Tips for Improving Flow
2 and 4	2 admires 4's depth, while 4 feels appreciated by 2's care.	Emotional understanding; 2 provides stability for 4's intensity.	2 may feel overwhelmed by 4's mood swings; 4 may find 2's positivity excessive.	Encourage open dialogue and mutual acceptance of emotional rhythms.
2 and 5	Type 2 draws out 5's emotions, while 5 brings intellectual depth to 2's world.	Complementary strengths; 2 nurtures, and 5 provides insight.	5 may find 2 intrusive; 2 may feel 5 is emotionally distant.	Respect personal boundaries and cultivate patience with differing needs.
2 and 6	Type 2 reassures 6, while 6 provides security and loyalty to 2.	Trust and mutual support; 6 admires 2's warmth.	6's anxiety may overwhelm 2; 2 may feel unreciprocated in emotional care.	Practice open communication and encourage independence.
2 and 7	2 enjoys 7's energy, while 7 appreciates 2's grounding care.	Fun and emotional connection; 2 nurtures while 7 inspires joy.	7 may appear unreliable; 2 may feel taken for granted.	Balance spontaneity with deeper emotional commitment.

Pair	Compatibility with Type 2	Strengths	Challenges	Tips for Improving Flow
2 and 8	2 softens 8's intensity, while 8 protects 2's vulnerability.	Strong and supportive; 2 nurtures while 8 empowers.	8's assertiveness may overwhelm 2; 2 may rely too heavily on 8 for validation.	Foster balanced power dynamics and mutual respect.
2 and 9	2 energizes 9, while 9 provides calm and steadiness for 2.	Peaceful and nurturing; 2 inspires action in 9.	Risk of complacency; 2 may feel underappreciate d.	Encourage active engagement and express mutual gratitude.

Type 3: The Achiever (Goal-driven, Ambitious, High-performing, Image-focused)

Driven and ambitious, Achievers focus on success and productivity, striving to excel and gain recognition. They are adaptable and highly motivated to accomplish their goals. Each type develops unique coping mechanisms—patterns of thought, emotion, and behavior formed to address their core fears and navigate life's challenges. For Achievers, this often involves prioritizing performance and image, masking vulnerability to maintain an outward appearance of success. While this fosters achievement, it can also lead to feelings of disconnection or burnout.

In the following sections, we explore how Type 3 typically operates in key areas of life—at work, home, and in relationships—revealing their characteristic strengths, vulnerabilities, and tendencies. We also examine this type's consciousness and plasticity challenges, identifying opportunities to enhance self-awareness, adaptability, and emotional intelligence. To support these efforts, we provide tailored *OPEN rituals*—PRIMAL, RATIONAL, and SPIRITUAL practices that nurture openness, balance, and personal growth.

Finally, the section concludes by unpacking compatibility dynamics, offering insights into how Type 3 relates to all other Enneagram types across various professional, familial, or romantic relationships.

The Gifts of Being an Enneagram 3

1. **Inspiration and Motivation:** Type 3s can inspire and motivate those around them. Their ambition and drive often catalyze action in others, helping them set goals and work towards achieving them. Type 3s naturally encourage others to reach their full potential by modeling determination and success, whether in leadership roles, friendships, or family settings.

2. **Goal-Oriented Focus and Efficiency:** Type 3s excel at setting and achieving clear goals. They are incredibly focused and results-driven, highly effective in work and personal projects.

Their natural talent for organization and time management allows them to streamline processes, prioritize tasks, and get things done efficiently and precisely.

3. **Adaptability and Versatility:** One of the key gifts of Type 3s is their adaptability. They are highly attuned to their environment's expectations and can easily adjust their approach or behavior to fit the situation. This makes them incredibly versatile and able to thrive in various settings, whether leading a team, connecting socially, or navigating family dynamics.

5. **Charisma and Emotional Intelligence:** Type 3s are naturally charismatic, and their emotional intelligence allows them to connect with others on a deep level when they choose to. Their ability to understand social cues and present themselves in ways that resonate with their audience makes them influential and likable. They are often seen as confident and competent, which draws people to them and enables them to lead effectively.

6. **Drive for Excellence and Improvement:** Type 3s strongly desire excellence in everything they do. They continually strive to be their best version, pushing the boundaries of what they can achieve. This constant drive for self-improvement ensures that they not only meet but often exceed expectations in their personal and professional lives.

The gifts of Type 3s lie in their ability to inspire, drive results, and adapt to different circumstances. Their charisma and emotional intelligence make them influential leaders and strong relationship-builders. Focusing on excellence and improvement, they set high standards for themselves and others, continuously motivating those around them to aim for success. By embracing these gifts while balancing their emotional needs, Type 3s can lead fulfilling and impactful lives.

Type 3 Coping Mechanisms

Type 3s must learn to cultivate self-worth independent of

external achievements to move beyond these coping mechanisms. By embracing vulnerability, allowing themselves to feel and express their emotions, and connecting with their authentic desires, Type 3s can develop a stronger sense of identity. Learning to appreciate themselves for who they are—not just what they accomplish—will allow them to form deeper relationships and experience more fulfilling, balanced lives.

1. **Over-reliance on External Validation**

   - **Coping Mechanism**:
     As children, Type 3s often learn to gain approval and love by excelling in tasks or meeting high expectations set by caregivers. They developed a habit of linking their self-worth to achievements, believing success was the key to love and acceptance.
   - **Adult Challenge**:
     In adulthood, this can lead to workaholism and prioritizing external recognition over personal fulfillment. Type 3s may struggle with feelings of inadequacy when they fail to meet their or others' high expectations, and they may avoid vulnerability to maintain their image of success.

2. **Emotional Disconnection**

   - **Coping Mechanism**:
     Type 3s learned to suppress emotions like fear or sadness that could disrupt their pursuit of excellence. They prioritized performance over authenticity, distancing themselves from emotions to stay focused on success.
   - **Adult Challenge**:
     As adults, this emotional suppression can make it difficult for Type 3s to connect deeply with others. They may avoid introspection or deny emotional struggles, fearing it could tarnish their image or slow their momentum.

3. **Perfectionism and Fear of Failure**

- **Coping Mechanism**:
  To avoid criticism or disappointment, Type 3s developed a perfectionist mindset. They worked tirelessly to succeed, fearing failure would diminish their value in the eyes of others.
- **Adult Challenge**:
  In adulthood, this fear of failure can lead to burnout or an inability to celebrate achievements. Type 3s may struggle with imposter syndrome, feeling their worth is tied to constant productivity.

4. Emotional Manipulation to Gain Connection

- **Coping Mechanism:**
  When Type 2s felt neglected or unloved, they may have learned to use subtle emotional manipulation—such as guilt or praise—to maintain closeness with others.
- **Adult Challenge:**
  In adulthood, this behavior can create unhealthy dynamics in relationships. Type 2s may unintentionally foster codependency or feel hurt when their efforts to secure connection are not reciprocated.

5. Difficulty Expressing Vulnerability

- **Coping Mechanism:**
  To maintain their role as helpers or nurturers, Type 2s often avoid showing vulnerability, believing they must be strong for others.
- **Adult Challenge:**
  Suppressing vulnerability can lead to a lack of emotional intimacy in relationships. Type 2s may struggle to share their struggles or ask for support, perpetuating feelings of isolation and unfulfilled emotional needs.

The gifts of Type 2s—empathy, generosity, emotional intelligence, and the ability to build deep connections—make them invaluable in every aspect of life. They bring love, support, and compassion wherever they go, nurturing the emotional and relational well-being of everyone they encounter. By fully embracing their gifts while learning to set healthy boundaries and care for themselves, Type 2s can experience true fulfillment and create lasting, meaningful change in their relationships and communities.

The challenges that arise from Type 2's coping mechanisms—such as overgiving, neglecting self-care, and seeking validation through service—can lead to emotional exhaustion, resentment, and a lack of fulfillment. By recognizing these tendencies and learning to set healthy boundaries, Type 2s can begin to balance their natural generosity with self-care, ultimately fostering healthier, more fulfilling relationships with them and others. The key for Type 2s is learning to love and care for themselves with the compassion they offer others.

Type 3 at Work

As an Enneagram Type 3, known as The Achiever, your primary focus is success, productivity, and recognition. You thrive in competitive environments where your achievements are noticed and rewarded. Type 3s are highly motivated, ambitious, and goal-oriented, which makes them excellent leaders and workers in any professional setting. Your natural charisma and ability to adapt to various roles ensure that you stand out and get the job done.

Strengths

⇒ **Goal-oriented:** Type 3s excel in setting and achieving ambitious goals, making them highly productive and efficient.
⇒ **Motivational:** Their confidence and energy inspire teams to strive for success and maintain a positive outlook.
⇒ **Adaptable:** They quickly adjust to changing circumstances, finding innovative ways to overcome challenges.
⇒ **Leadership Skills:** Type 3s are natural leaders, excelling in roles that require vision, strategic planning, and execution.

⇒ **Results-driven:** They are focused on delivering measurable outcomes, which makes them valuable in competitive environments.

Challenges

⇒ **Overworking:** Their drive for success can lead to burnout or neglect of work-life balance.

⇒ **Perfectionism:** They may prioritize image and results over collaboration, leading to strained relationships.

⇒ **Difficulty Accepting Failure:** A strong aversion to failure can result in risk avoidance or defensiveness.

⇒ **Overemphasis on External Validation:** They might base their self-worth on recognition and accolades, impacting their authenticity.

⇒ **Delegation Issues:** Type 3s may struggle to trust others to perform tasks, preferring to take on too much themselves.

Type 3 at Home

At home, Type 3s often bring their productivity-driven mindset into their personal lives. They are goal-setters and organizers, ensuring their household is efficient and runs smoothly. They may also value presenting their home and family in a way that reflects their success, desiring admiration from others.

Strengths

⇒ **Driven to Provide**: Type 3s are highly motivated to meet their family's needs, often creating a secure and comfortable environment.

⇒ **Supportive of Goals**: Encourages loved ones to pursue their dreams and aspirations, providing inspiration and motivation.

⇒ **Organized and Efficient**: Their focus on productivity helps them manage household responsibilities effectively, often keeping things running smoothly.

⇒ **Charismatic Presence**: Their charm and confidence can make them engaging and enjoyable companions for family members.

⇒ **Solution-oriented**: Quickly addressing challenges or conflicts, finding practical and effective resolutions to problems.

⇒ **Encourages Growth**: Strives to cultivate an atmosphere where loved ones feel empowered to improve and succeed.

Challenges

⇒ **Work-life Imbalance**: Their intense focus on achievement and external success can lead to neglecting personal relationships and quality time.

⇒ **Overemphasis on Appearances**: May prioritize how things look to outsiders, creating pressure for the family to meet specific standards.

⇒ **Difficulty Showing Vulnerability**: Prefers to project confidence and competence, which can prevent deeper emotional connections.

⇒ **Conditional Support**: Might unconsciously tie love or approval to performance, leading family members to feel judged or unaccepted for who they are.

⇒ **Impatience with Emotions**: Can struggle to navigate or validate emotions that don't align with their practical, goal-oriented mindset.

⇒ **Stress Under Pressure**: Their fear of failure or imperfection can make them irritable or overly controlling in stressful situations.

⇒ **Neglecting Authenticity**: May unintentionally downplay their true feelings to maintain a positive or "successful" image within the family.

Type 3 in Relationship

In relationships, Type 3s are charming, driven, and supportive. They are often admired for their ambition and ability to inspire others to reach their full potential. However, Type 3s can sometimes prioritize their success and recognition over emotional intimacy, which may create distance in relationships.

Strengths

⇒ **Supportive Partner**: Type 3s encourage their partners to pursue their goals and aspirations, offering motivation and practical help.
⇒ **Goal-Oriented Love**: They work hard to build a successful and fulfilling relationship, often striving for an ideal partnership.
⇒ **Charismatic and Engaging**: Their charm and confidence make them attractive and enjoyable companions.
⇒ **Resourceful and Capable**: They bring organization, efficiency, and a proactive attitude to relationships, solving problems quickly and effectively.
⇒ **Hardworking and Reliable**: Committed to ensuring the relationship thrives, they often invest significant effort to provide emotional and material stability.
⇒ **Inspiring Growth**: They challenge their partners to grow and achieve their full potential, creating a dynamic and aspirational bond.
⇒ **Polished and Presentable**: They often ensure that shared experiences and partnerships appear polished and harmonious, enhancing their shared social image.

Challenges

⇒ **Over-focus on Image**: They may prioritize how the relationship appears to others over its authenticity or emotional depth.
⇒ **Difficulty Showing Vulnerability**: They often avoid exposing their insecurities or weaknesses, which can hinder emotional intimacy.

⇒ **Workaholic Tendencies**: Their drive for success can leave partners feeling neglected or undervalued due to insufficient time together.

⇒ **Conditional Love**: May unconsciously tie their affection or approval to their partner's achievements or outward success.

⇒ **Impatience with Emotions**: They might struggle to fully engage with or validate their partner's emotional needs, focusing instead on solutions or goals.

⇒ **Fear of Failure**: This can manifest as difficulty addressing conflicts or admitting to mistakes, creating tension in the relationship.

⇒ **Over-competitiveness**: They may compare their success to their partner's, potentially creating feelings of rivalry or imbalance.

Type 3 Consciousness Challenges

Type 3s, the driven and success-oriented "Achievers," often prioritize external validation and accomplishments over self-reflection. Their focus on image and productivity can disconnect them from their authentic feelings and inner values. This tendency to align identity with success may hinder deeper self-awareness and cultivating genuine emotional presence.

1. Over-Identification with Image

Type 3s tend to overidentify with their external image and success, making it difficult to connect with their true, authentic selves. They often prioritize how others perceive them, maintaining a polished and competent exterior. This makes it challenging for Type 3s to engage in deep self-reflection or inner emotional work, as they can become disconnected from their inner feelings and desires in favor of external approval and recognition.

2. Avoidance of Vulnerability

Type 3s frequently avoid vulnerability, as it threatens the successful and confident image they work so hard to maintain. Vulnerability requires embracing weaknesses, fears, and insecurities—elements Type 3s prefer to suppress. Their fear of appearing emotionally weak or imperfect hinders their ability to develop deeper consciousness. True self-awareness requires acknowledging and embracing vulnerabilities, but Type 3s often resist this emotional depth to maintain their image of success.

3. **Reluctance to Slow Down and Reflect**

Constantly pursuing the next achievement, Type 3s often struggle to slow down and reflect. Their focus on productivity and efficiency leaves little room for introspection or mindfulness. Reflection is key to raising consciousness, but Type 3s may see it as unproductive or even uncomfortable. Without moments of stillness and self-reflection, Type 3s miss the opportunity to explore their deeper motivations, feelings, and emotional needs.

4. **Fear of Failure**

Type 3s' fear of failure can block their consciousness development. They strive for success and perfection, often viewing failure as a threat to their identity. This fear keeps them from taking risks or engaging in activities where they might not excel. Consciousness growth requires learning from mistakes, but Type 3s may avoid confronting their fear of failure, limiting their emotional and personal development.

Type 3: Plasticity Challenges

Neuroplasticity refers to the brain's ability to reorganize itself by forming new neural connections throughout life. It allows people to change habits, beliefs, and emotional responses through new experiences and learning.

1. **Rigid Focus on Success**

Type 3s are driven by a singular focus on achieving success, which can become rigid and inflexible. Their determination to accomplish set goals limits their openness to new perspectives, ideas, and ways of thinking, which are essential for developing plasticity. While this focus helps them succeed in familiar areas, it may prevent them from embracing the change and adaptability required to form new neural pathways and explore new approaches.

2. Over-reliance on External Validation

Type 3s are heavily reliant on external validation for their sense of self-worth. They often seek praise and recognition from others, which narrows their focus to activities that bring immediate rewards. This external focus can limit their ability to engage in long-term personal development or practices that promote plasticity, such as mindfulness or emotional growth, because they prioritize what will gain approval over what will foster internal change.

3. Avoidance of Vulnerability

Vulnerability is essential for plasticity, allowing individuals to step outside their comfort zones, learn from mistakes, and embrace new experiences. Type 3s, however, tend to avoid vulnerability to protect their image of success. This avoidance prevents them from engaging in new challenges or unfamiliar situations where failure might occur. By sticking to familiar routines or activities where they know they will succeed, Type 3s miss opportunities to build resilience and form new neural pathways.

4. Struggle with Authenticity

Type 3s often focus on maintaining an image of success and competence, which can create a disconnect between their true selves and their public persona. Authenticity is crucial for brain plasticity, as genuine emotional engagement helps create lasting behavioral and cognitive changes. The pressure to perform and meet external expectations can keep Type 3s from exploring who they are, limiting the development of new, more authentic emotional responses and behaviors.

Type 3: Summary of Challenges

Challenge	Impact on Consciousness	Impact on Plasticity
Over-Identification with Image	Prevents self-awareness by focusing on external approval over inner needs.	Limits flexibility by focusing only on activities that maintain their image.
Avoidance of Vulnerability	It blocks emotional depth and prevents true self-reflection and personal growth.	Avoids new experiences that require risk, preventing the formation of new neural pathways.
Reluctance to Slow Down and Reflect	Lack of introspection hinders deeper emotional awareness and self-exploration.	It prevents the development of new habits or insights that come from mindful reflection.
Fear of Failure	Limits growth by avoiding risks and blocking learning from mistakes.	Inhibits adaptability by sticking to familiar tasks to avoid failure.
Rigid Focus on Success	Narrow focus on achievement restricts inner exploration and emotional balance.	Limits openness to change and new approaches, blocking mental flexibility.
Over-Reliance on External Validation	Dependence on external praise prevents internal self-worth development.	Focuses on short-term approval, avoiding practices that foster long-term brain growth.
Struggle with Authenticity	Creates disconnect between true self and public persona, limiting self-awareness.	Limits plasticity by blocking emotionally authentic learning and growth experiences.

Type 3: PRIMAL Rituals Recommended

For Type 3s, primal rituals slow the pace of constant striving, helping them reconnect with their body and emotions. These practices encourage vulnerability by allowing them to step away from performance-driven behavior and focus on inner balance.

Open up to Your Body

⇒ **Why**: Encourages Type 3s to reconnect with their physical self and focus on well-being rather than external performance.
⇒ **Frequency**: Daily.

Open up to Your Breath

⇒ **Why**: Breathwork helps Type 3s manage stress and cultivates mindfulness, redirecting their focus inward.
⇒ **Frequency**: Daily for 5-10 minutes.

Open up to Nature

⇒ **Why**: Nature immersion reduces performance pressure and fosters inner calm and self-reflection.
⇒ **Frequency**: Weekly.

Open up to Your Sleep

⇒ **Why**: Prioritizing restorative sleep helps Type 3s recharge emotionally and physically, preventing burnout.
⇒ **Frequency**: Nightly.

Open up to Your Food

⇒ **Why**: Mindful eating encourages Type 3s to slow down, focusing on nourishment rather than productivity.
⇒ **Frequency**: Each meal.

Type 3: RATIONAL Rituals Recommended

RATIONAL rituals guide Type 3s toward self-reflection, reducing their dependence on external validation. These practices help them embrace authenticity, fostering emotional vulnerability and a more meaningful connection with others.

Open up to Learning

⇒ **Why**: Expanding knowledge encourages Type 3s to explore personal growth beyond achievement and external validation.
⇒ **Frequency**: Weekly.

Open up to Evocative Music

⇒ **Why**: Evocative music helps Type 3s process emotions they may suppress in their pursuit of success.
⇒ **Frequency**: Weekly.

Open up to Cooking

⇒ **Why**: Cooking as a creative and mindful activity helps Type 3s reconnect with themselves and enjoy the process without focusing on outcomes.
⇒ **Frequency**: Several times a week.

Open up to Traveling

⇒ **Why**: Traveling exposes Type 3s to new experiences and perspectives, shifting their focus away from constant productivity.
⇒ **Frequency**: Periodically, as opportunities arise.

Open up to Laughter

⇒ **Why**: Humor and lightheartedness help Type 3s let go of performance pressure and embrace joy in the present moment.
⇒ **Frequency**: As often as possible.

Type 3: SPIRITUAL Rituals Recommended

SPIRITUAL rituals for Type 3s emphasize releasing attachment to success and status. Through meditative or creative activities, they can reconnect with their true essence and explore fulfillment beyond achievement.

Open up to Meditation

⇒ **Why**: Meditation fosters inner peace and self-awareness, allowing Type 3s to connect with their authentic self.
⇒ **Frequency**: Daily, 10-15 minutes.

Open up to Prayers

⇒ **Why**: Prayers provide Type 3s with a sense of purpose and gratitude beyond their achievements.
⇒ **Frequency**: As needed, especially in moments of self-doubt.

Open up to Stargazing

⇒ **Why**: Stargazing reminds Type 3s of the vastness of life, helping them shift focus from external achievements to inner wonder.
⇒ **Frequency**: Monthly or during clear nights.

Open up to Your Dreams

⇒ **Why**: Reflecting on dreams helps Type 3s uncover subconscious thoughts and reconnect with their inner emotional world.

⇒ **Frequency**: Nightly journaling upon waking.

Open up to Mantras

⇒ **Why**: Repeating affirmations helps Type 3s challenge the need for external validation and build internal confidence.
⇒ **Frequency**: Daily.

Type 3: Compatibility With Other Types

Enneagram Type 3, known as **The Achiever**, is driven, goal-oriented, and focused on success. Type 3s excel in motivation and efficiency in personal or professional relationships but may struggle with emotional depth or vulnerability. Below is an analysis of how Type 3 interacts with each of the other nine types, highlighting areas of compatibility and potential friction.

Type 1 (The Reformer)

•**What Works**: Type 1 and Type 3 focus on improving and achieving high standards. Type 3's ambition pairs well with Type 1's sense of responsibility, making them an effective team.
•**Potential Friction**: Type 1's need for perfection may clash with Type 3's tendency to prioritize image over substance. Type 1s may see Type 3s as too focused on appearances, while Type 3s may feel judged by Type 1's high moral expectations.

Type 2 (The Helper)

•**What Works**: Type 2's nurturing nature complements Type 3's drive for success. Type 3 appreciates Type 2's emotional support and values Type 2's encouragement of personal growth.
•**Potential Friction**: Type 3's focus on goals may cause them to overlook Type 2's emotional needs. Type 2 may feel underappreciated if Type 3 prioritizes work or external validation over the relationship.

Type 3 (The Achiever)

•**What Works**: Two Type 3s create a dynamic, high-achieving partnership where both push each other to succeed. They can collaborate effectively on projects and share a drive for success.
•**Potential Friction**: The relationship may become too competitive, with both focusing more on accomplishments than on emotional connection. They may struggle to prioritize vulnerability and depth in their interactions.

Type 4 (The Individualist)

•**What Works**: Type 4's emotional depth balances Type 3's external focus on success. Type 4 may help Type 3 access deeper feelings, while Type 3 can motivate Type 4 to pursue their creative goals.
•**Potential Friction**: Type 4's focus on authenticity and emotional exploration may clash with Type 3's tendency to suppress emotions to maintain an image. Type 4 may feel that Type 3 is superficial, while Type 3 may see Type 4 as overly dramatic.

Type 5 (The Investigator)

•**What Works**: Type 5's analytical approach and Type 3's execution-focused mindset can complement each other well, making them a strong team in work or intellectual pursuits.
•**Potential Friction**: Type 3's need for recognition may clash with Type 5's desire for independence and privacy. Type 5 may find Type 3's emphasis on external validation draining, while Type 3 may feel that Type 5 is too detached or unemotional.

Type 6 (The Loyalist)

•**What Works**: Type 6's loyalty and focus on security align well with Type 3's ambition. Type 6 can help ground Type 3, offering support and helping them think through potential risks.
•**Potential Friction**: Type 6's anxiety and tendency to second-guess decisions may frustrate Type 3, who prefers decisive action. Type 3 may also appear too focused on achievement, leaving Type 6 feeling

insecure or neglected in the relationship.

Type 7 (The Enthusiast)

•**What Works**: Type 7's optimism and love for adventure pair well with Type 3's ambitious energy. Both types enjoy staying active and pursuing new experiences, creating a fun and dynamic partnership.
•**Potential Friction**: Type 3's goal-driven mindset may conflict with Type 7's tendency to avoid commitment or follow-through. Type 7 might see Type 3 as too focused on work, while Type 3 might view Type 7 as too unfocused or irresponsible.

Type 8 (The Challenger)

•**What Works**: Type 8's boldness and assertiveness complement Type 3's ambition, creating a powerful and productive partnership. Both types appreciate strength and decisiveness.
•**Potential Friction**: Type 8's directness and desire for control may clash with Type 3's adaptability and image-conscious behavior. Type 8 may view Type 3 as manipulative or inauthentic, while Type 3 could feel overpowered or controlled by Type 8's intensity.

Type 9 (The Peacemaker)

•**What Works**: Type 9's calm, easy-going nature balances Type 3's high energy and ambition. Type 9 provides emotional stability, while Type 3 offers motivation and direction.
•**Potential Friction**: Type 9's tendency to avoid conflict and decisions may frustrate Type 3, who wants to move forward quickly. Type 9 might feel overlooked or pushed aside if Type 3 is too focused on external goals rather than the relationship.

Type 3 generally thrives in relationships where their ambition and drive are appreciated, and their partner complements their focus on success. However, friction can arise when Type 3s prioritize achievements or external validation over emotional depth and vulnerability. Balancing the drive for success with emotional awareness is key for Type 3 in building harmonious relationships.

Type 3: Compatibility Table

Pair	Compatibility with Type 3	Strengths	Challenges	Tips for Improving Flow
3 and 1	Type 3's drive for success complements Type 1's dedication to integrity and perfection.	Together, they set high standards and achieve meaningful goals.	1 may view 3 as overly image-focused, while 3 may feel criticized by 1's high standards.	Balance ambition with authenticity and focus on shared values.
3 and 2	Type 2's warmth supports 3's aspirations, while 3 admires 2's emotional depth.	Inspiring partnership; 3 drives forward while 2 nurtures the connection.	2 may feel neglected if 3 prioritizes success over the relationship; 3 may feel smothered by 2's care.	Encourage mutual appreciation and open communication about needs.
3 and 3	Both thrive on achievement and enjoy mutual admiration for each other's successes.	High energy and shared ambition make them a dynamic pair.	Risk of competition and burnout due to relentless focus on success.	Cultivate deeper emotional connection beyond shared achievements.

Pair	Compatibility with Type 3	Strengths	Challenges	Tips for Improving Flow
3 and 4	3's practicality balances 4's emotional depth, creating a dynamic balance of action and introspection.	Together, they combine vision and execution for creative pursuits.	3 may dismiss 4's emotional intensity, while 4 may see 3 as inauthentic.	Embrace vulnerability and value each other's unique strengths.
3 and 5	3's action-oriented mindset complements 5's intellectual depth and planning.	Together, they combine knowledge with execution to achieve big-picture goals.	5 may find 3 too focused on external validation, while 3 may find 5 overly withdrawn.	Foster patience and appreciate differing approaches to problem-solving.
3 and 6	6 provides loyalty and caution to balance 3's risk-taking and drive.	They form a secure partnership, with 6 grounding 3's ambitious pursuits.	6's anxiety may frustrate 3, while 3's fast pace may overwhelm 6.	Build trust and align on shared goals to reduce stress and friction
3 and 7	Both love excitement and share a high-energy approach to life.	They enjoy fun, adventure, and tackling challenges together.	3's focus on achievement may clash with 7's need for freedom & spontaneity	Balance productivity with moments of unstructured joy and connection.

Pair	Compatibility with Type 3	Strengths	Challenges	Tips for Improving Flow
3 and 8	8's strength and decisiveness complement 3's drive for success, creating a powerful duo.	Together, they are highly goal-oriented and capable of significant achievements.	8 may overpower 3, while 3 may feel pressured to prove themselves constantly.	Practice mutual respect and embrace vulnerability to deepen the connection.
3 and 9	3's ambition motivates 9 to take action, while 9's calm demeanor helps 3 relax.	They create a balanced dynamic of action and stability.	3 may feel 9 lacks ambition, while 9 may feel pushed by 3's intensity.	Encourage mutual understanding and align on shared aspirations.

Type 4: The Individualist (Emotional, introspective, creative, moody)

Authentic and introspective Individualists are deeply attuned to their emotions and seek meaning through self-expression and uniqueness. They often prioritize depth and creativity. Each type develops unique coping mechanisms—patterns of thought, emotion, and behavior formed to address their core fears and navigate life's challenges. For Individualists, this often involves dwelling on their feelings and focusing on what feels missing in their lives.

While this fosters depth and creativity, it can also lead to cycles of melancholy or isolation. In the following sections, we explore how Type 4 typically operates in key areas of life—at work, home, and in relationships—revealing its characteristic strengths, vulnerabilities, and tendencies. We also examine this type's consciousness and plasticity challenges, identifying opportunities to enhance self-awareness, adaptability, and emotional intelligence.

To support these efforts, we provide tailored *OPEN rituals*—PRIMAL, RATIONAL, and SPIRITUAL that nurture openness, balance, and personal growth. Finally, the section concludes by unpacking compatibility dynamics, offering insights into how Type 4 relates to all other Enneagram types across various professional, familial, or romantic relationships.

The Gifts of Being an Enneagram 4

1. **Creativity and Originality:** Type 4s are inherently creative, often expressing their thoughts and emotions through art, music, writing, or other forms of expression. They thrive in environments where they can use their imagination to bring something original. This gift extends beyond traditional arts; even in everyday situations, Fours is known for approaching problems with innovative solutions.

2. **Emotional Depth and Insight**: Fours possess a rich emotional landscape and can easily connect with the deeper emotions of others. This gives them a heightened sense of empathy and an ability to offer profound emotional support to

those around them. Their understanding of their emotions makes them capable of having transformative conversations and helping others navigate their feelings.

3. **Authenticity and Self-Awareness:** One of the core desires of Type 4 is to live authentically. They seek to be true to themselves and resist conforming to societal expectations if it means sacrificing their individuality. This quality often leads others to admire their courage to live authentically and vulnerably, inspiring those around them to do the same.

4. **Empathy and Compassion:** Type 4s' deep emotional wellspring makes them naturally compassionate individuals. They are often drawn to help those suffering because they can intimately understand pain, sadness, and longing. Their empathy also makes them excellent listeners and supportive friends, partners, and colleagues.

5. **Passion and Intensity:** Fours are known for their passion. When they commit to something they care about—a cause, a person, or a project—they give it their all. Their intensity can motivate and inspire those around them, as they can channel this energy into creating beautiful, meaningful things.

6. **Intuition and Depth of Perception**: Fours are highly sensitive to their environment and others' emotions. They often intuitively sense underlying currents in social situations and can perceive meanings, symbols, and connections others might overlook. This makes them great at uncovering hidden patterns or understanding the emotional subtext in conversations.

Type 4 Coping Mechanisms

Type 4s, the Individualists, often cope by intensifying their emotions and seeking uniqueness to feel significant. They may dwell on feelings of inadequacy or loss, channeling their energy into self-expression or creative pursuits. This mechanism helps them manage their fear of being ordinary but can also deepen their sense of disconnection or longing if left unchecked.

1. **Emotional Intensity and Mood Swings**

    - **Coping Mechanism**: As children, Type 4s may have experienced deep emotional highs and lows, often feeling misunderstood or different from those around them. They coped by channeling their emotions into creative outlets or retreating into their inner world to find meaning and solace.
    - **Adult Challenge**: In adulthood, this can lead to mood swings that make maintaining consistency in relationships and tasks difficult. Type 4s may feel overwhelmed by their emotions, struggling to stay grounded and present when emotions become intense.

2. **Self-absorption and Over-Identification with Feelings**

    - **Coping Mechanism**: Type 4s may have coped by overly identifying with their feelings, believing that their emotions defined who they were. This tendency gave them a sense of identity and led them to focus excessively on their inner world.
    - **Adult Challenge**: This can result in self-absorption in adulthood, making it hard to connect with others or see situations objectively. Type 4s may dwell on feelings of inadequacy or rejection, often struggling to shift focus toward solutions or broader perspectives.

3. **Envy and Comparison**

    - **Coping Mechanism**: Feeling like something was missing in their lives, Type 4s coped by comparing themselves to others, admiring what others had that they perceived they lacked. This gave them a sense of what they longed for and deepened their feelings of inadequacy.
    - **Adult Challenge**: This tendency toward envy can intensify longing and dissatisfaction in adulthood. Type 4s may

compare their lives unfavorably to others, struggling to appreciate their achievements or blessings.

4. **Fear of Abandonment and Rejection**

    - **Coping Mechanism**: Type 4s often feared being abandoned or rejected due to feeling different or misunderstood. They coped by withdrawing emotionally or creating an idealized version of themselves to gain acceptance.
    - **Adult Challenge**: In adulthood, this fear can make them sensitive to perceived rejection, leading them to withdraw or self-isolate when they feel vulnerable. They may struggle to open up fully, fearing their emotional intensity will push others away.

5. **Idealization and Disappointment**

    - **Coping Mechanism**: To cope with reality's imperfections, Type 4s often idealized people, situations, or relationships, creating an inner vision of how things should be. This allowed them to escape from disappointments or painful realities.
    - **Adult Challenge**: This pattern can lead to chronic disappointment in adulthood when reality fails to meet their idealized expectations. Type 4s may struggle with sustaining long-term contentment, as they may always feel something is missing.

Type 4 at Work

Enneagram 4s thrive in environments that allow for creativity and self-expression. They prefer roles where they can bring their unique vision or creative input to the table, often excelling in fields like the arts, design, writing, or roles that require deep emotional intelligence. However, they may struggle in rigid, highly structured environments that stifle individuality. Type 4s might occasionally compare their contributions to others in the workplace, leading to inadequacy or frustration if they feel their unique talents aren't appreciated.

Strengths

⇒ **Creativity and Innovation**: Type 4s excel in roles requiring out-of-the-box thinking. They bring a strong artistic vision and originality to their work. They thrive in fields that allow for self-expression, such as design, writing, or problem-solving.

⇒ **Emotional Insight**: They have a deep understanding of human emotions, which makes them empathetic colleagues and valuable contributors to team dynamics. This emotional depth helps them connect profoundly with clients or audiences.

⇒ **Authenticity and Individuality**: Type 4s strive to create meaningful and authentic work. They bring passion and personal investment to projects, often inspiring others.

⇒ **Attention to Aesthetics and Quality**: They have a natural talent for recognizing beauty and detail, making them strong contributors to projects requiring a refined touch.

Challenges

⇒ **Mood and Sensitivity**: Their emotional intensity can make them moody or overly sensitive to criticism, impacting their resilience in high-pressure or collaborative environments.

⇒ **Tendency to Feel Misunderstood**: Type 4s may isolate themselves if they feel undervalued or misaligned with the workplace culture, leading to disengagement.

⇒ **Idealism and Comparison**: They may struggle with dissatisfaction if their work doesn't align with their idealized vision or if they compare themselves unfavorably to others.

⇒ **Difficulty with Routine Tasks**: Type 4s prefer creative, dynamic work and may find repetitive or administrative tasks draining or uninspiring.

Type 4 at Home

Type 4s value deep, meaningful connections at home with their loved ones. They often strive to create an environment that reflects their personal style and emotional landscape, curating authentic spaces. They seek to be understood deeply, often yearning for intimate and sincere conversations. However, their emotional highs and lows can sometimes make them feel disconnected from others, especially if they sense that their feelings are not fully acknowledged.

Strengths

⇒ **Emotional Depth**: Type 4s bring a deep emotional understanding to their home environment. They can empathize with family members and provide a comforting space for expressing feelings.

⇒ **Creative and Aesthetic Sensibilities**: They often create a unique and beautiful home environment, as they value individuality and artistic expression.

⇒ **Passion for Authenticity**: Type 4s encourage honest communication and authentic relationships, fostering deep connections with their loved ones.

Challenges

⇒ **Moodiness**: Their emotional intensity can sometimes lead to mood swings, creating unpredictability or tension in family dynamics.

⇒ **Sensitivity to Feeling Misunderstood**: Type 4s strongly desire to be seen and understood. They may withdraw or become overly self-focused if they feel ignored or undervalued.

⇒ **Tendency Toward Comparison**: They may compare their family dynamics or home life to idealized standards, leading to dissatisfaction or a sense of inadequacy

Type 4 in Relationship

Type 4s seek profound emotional connections and long for partners who understand their depth. They are often introspective and passionate lovers, offering a level of emotional intensity that can be enriching for their partners. However, they may also struggle with envy, particularly if they perceive others as having more fulfilling lives or relationships. Sometimes, they may pull away emotionally if they feel their uniqueness isn't appreciated.

Strengths

⇒ **Mood Swings**: Their emotional intensity can lead to moodiness, which might be overwhelming or confusing for their partner.

⇒ **Desire to Feel Special**: Type 4s strongly need to feel unique and understood. They may feel neglected or withdrawn if their partner doesn't meet this need.

⇒ **Idealism and Comparison**: People may idealize relationships or compare their partners to an unrealistic standard, leading to disappointment or dissatisfaction.

⇒ **Sensitivity to Perceived Rejection**: Type 4s can be highly sensitive to any signs of disconnection or misunderstanding, which may cause them to retreat or become overly self-focused.

Challenges

⇒ **Foster Open Communication**: Reassure Type 4s through honest conversations and emotional validation to help them feel seen and understood.

⇒ **Encourage Emotional Stability**: Support practices like mindfulness or journaling to help them regulate their intense emotions.

⇒ **Celebrate Their Uniqueness**: Acknowledge their individuality and contributions to the relationship to affirm their sense of specialness.

⇒ **Ground Idealism with Realism**: Encourage gratitude

and appreciation for the present moment, balancing their idealism with a focus on the positive aspects of the relationship.

Type 4 Consciousness Challenges

Type 4s, the introspective and emotionally rich "Individualists," often become deeply absorbed in their inner world and personal emotions. While this depth can be a strength, it may lead to over-identification with feelings of inadequacy or longing. Their focus on what is missing can hinder their ability to stay present and appreciate the completeness of their current experiences, challenging their journey toward higher consciousness.

1. **Over-Identification with Emotions**

Type 4s often struggle to overidentify with their emotions, making it challenging to maintain a state of objective self-awareness. Instead of seeing feelings as temporary states, they may become engulfed by them, believing they *are* their feelings. This over-identification limits their ability to step back, observe their experiences from a higher level of consciousness, and avoid emotional reactivity.

1. **Self-Absorption and Limited Perspective**

Due to their introspective nature, Fours can become overly focused on their inner world. This intense self-focus can prevent them from achieving higher states of consciousness, where they can recognize broader perspectives beyond their personal experience. This myopia can narrow awareness, making it harder for them to connect with others or embrace a more universal understanding of their experiences.

2. **Resistance to Present Moment Awareness**

Fours often dwell in their emotions or fantasize about idealized futures, making it difficult to remain present. Maintaining consciousness in the present moment is key to personal growth, yet

Fours' focus on what's missing or their yearning for something more can draw them away from mindfulness. This lack of presence prevents them from experiencing the full depth of their current experiences, limiting their potential for transformation.

3. **Fear of Being Misunderstood**

A common challenge for Type 4s is the fear of being misunderstood, which can block their access to higher consciousness. This fear may cause them to self-protect by withdrawing or amplifying their individuality in ways that reinforce their sense of separation. Instead of embracing higher consciousness's interconnectedness, they may remain stuck in their feelings of being different, increasing their sense of isolation.

4. **Idealization of Consciousness States**

Fours are prone to idealizing certain emotional or spiritual states, which can lead them to chase after peak experiences rather than appreciating ordinary moments of awareness. This idealization of consciousness can cause frustration when their daily life doesn't live up to their imagined spiritual ideals, making it harder for them to appreciate their current state of being and work steadily on their self-development.

Type 4: Plasticity Challenges

Neuroplasticity refers to the brain's ability to reorganize itself by forming new neural connections throughout life. It allows people to change habits, beliefs, and emotional responses through new experiences and learning.

1. **Rigidity in Emotional Patterns**

Type 4s may struggle with emotional rigidity, especially regarding chronic feelings of sadness, longing, or envy. This emotional inflexibility limits their plasticity—the brain's ability to form new neural pathways and adapt to change. Fours may find it difficult to shift out of these recurring emotional patterns, reinforcing a fixed mindset that blocks their growth and reduces their capacity to experience new ways of thinking and feeling.

2. Difficulty Letting Go of Past Stories

Fours often hold onto past hurts or emotional narratives that define their identity. This attachment to personal stories, especially those centered on pain or rejection, limits their ability to rewire their brain for new possibilities. The more they cling to these narratives, the harder it becomes for them to adopt new thought patterns or embrace change, reducing their brain plasticity.

3. Resistance to Cognitive Reappraisal

A key mechanism for fostering brain plasticity is the ability to reappraise situations, cognitively seeing events from new perspectives. Fours may resist this process because they are deeply attached to their initial emotional responses. Their preference for emotional intensity over rational analysis can inhibit the cognitive flexibility needed to reshape their thoughts and reactions, limiting their capacity for mental adaptation.

4. Fixed Sense of Identity

Type 4s often define themselves through emotional depth and uniqueness, leading to a fixed sense of identity. This self-concept can become rigid, making it difficult for them to adapt to new experiences or challenges. Plasticity requires openness to change, but the tendency to over-identify with a specific identity can inhibit a Four's ability to evolve and develop new aspects of themselves.

5. Reluctance to Engage in New Habits

Brain plasticity is greatly enhanced through the practice of new habits, but Fours may be reluctant to engage in behaviors that they feel don't align with their emotional state. For instance, adopting routines like mindfulness, regular physical exercise, or even cognitive behavioral techniques might seem too mundane or counter to their emotional inclinations, preventing them from rewiring their brain for greater flexibility and resilience.

Type 4: Summary of Challenges

Challenge	Impact on Consciousness	Impact on Plasticity
Over-Identification with Emotions	Limits ability to observe emotions objectively, leading to emotional reactivity and reduced self-awareness.	Creates emotional rigidity, making it difficult to adapt to new emotional patterns.
Self-Absorption and Limited Perspective	Narrows awareness, limiting a broader perspective and making it hard to connect with others or higher consciousness.	Restricts openness to new experiences, reducing the brain's ability to form new neural pathways.
Resistance to Present Moment Awareness	Hinders mindfulness and the ability to stay grounded in the present, leading to constant emotional fluctuations.	Inhibits the development of adaptive habits that foster change and new cognitive processes.
Fear of Being Misunderstood	Reinforces feelings of separation, blocking a sense of interconnectedness and hindering the expansion of awareness.	Prevents new social interactions or experiences that challenge existing emotional and behavioral patterns.
Idealization of Consciousness States	Causes frustration when real-life experiences don't match ideals, leading to dissatisfaction with current consciousness.	Limits plasticity by encouraging avoidance of ordinary experiences necessary for building adaptive resilience.

Challenge	Impact on Consciousness	Impact on Plasticity
Rigidity in Emotional Patterns	Keeps individuals stuck in repetitive emotional states, preventing them from evolving emotionally and spiritually.	Reduces brain flexibility by reinforcing the same emotional pathways over time, blocking new patterns.
Difficulty Letting Go of Past Stories	Anchors consciousness in past pain, limiting awareness of the present and potential for growth.	Reduces the ability to rewire the brain for positive change, keeping old emotional circuits active.
Resistance to Cognitive Reappraisal	Prevents reevaluation of emotional responses, keeping individuals in fixed emotional states rather than adapting cognitively.	Inhibits brain flexibility needed to create new cognitive and emotional responses to changing situations.
Fixed Sense of Identity	Limits self-exploration and the ability to evolve beyond a specific emotional identity, hindering higher consciousness.	Prevents openness to new experiences, reducing the ability to develop new behaviors and emotional responses.
Reluctance to Engage in New Habits	Blocks growth in consciousness by avoiding practical tools like mindfulness or meditation that expand awareness.	Limits brain plasticity by resisting the formation of new habits that create more adaptable brain circuits.

Type 4: PRIMAL Rituals Recommended

Grounding rituals for Type 4s anchor them in the present, softening their tendency to dwell on emotions and intensify their feelings. These practices help them embrace vulnerability by reducing the focus on self and fostering physical connection.

Open up to Your Breath

⇒ **Why**: Helps Type 4s manage emotional intensity by grounding their feelings through focused breathing. Breathwork fosters mindfulness and balance.

⇒ **Frequency**: Daily for 5-10 minutes.

Open up to Your Body

⇒ **Why**: Encourages Type 4s to reconnect with their physical sensations, grounding them in the present and providing an outlet for emotional energy.

⇒ **Frequency**: Daily through yoga, stretching, or mindful movement.

Open up to Your Sleep

⇒ **Why**: Supports emotional regulation by ensuring restorative rest, helping Type 4s manage mood fluctuations more effectively.

⇒ **Frequency**: Prioritize 7-8 hours of sleep nightly.

Open up to Nature

⇒ **Why**: Provides Type 4s with an opportunity to find peace, inspiration, and grounding by connecting with the natural world.

⇒ **Frequency**: Weekly for at least 30 minutes.

Type 4: RATIONAL Rituals Recommended

RATIONAL rituals encourage Type 4s to balance their introspection with clarity and objectivity. These practices reduce emotional reactivity and invite a deeper connection with others by gently challenging their focus on uniqueness.

Open up to Evocative Music

⇒ **Why**: Resonates with Type 4s' emotional depth, offering a way to process and release emotions constructively.
⇒ **Frequency**: Weekly or as needed for emotional release.

Open up to Uplifting Movies

⇒ **Why**: Offers an outlet for emotional reflection and inspires hope, balancing Type 4s' introspective tendencies.
⇒ **Frequency**: Weekly or as desired.

Open up to Art

⇒ **Why**: Allows Type 4s to explore and express complex emotions through visual or creative practices, fostering self-expression.
⇒ **Frequency**: Weekly or as needed for creative expression.

Type 4: SPIRITUAL Rituals Recommended

SPIRITUAL rituals for Type 4s nurture a sense of belonging and release the ego's need for identity through uniqueness. They can cultivate inner peace and a connection to universal harmony through prayer or mindfulness.

Open up to Meditation

⇒ **Why:** It encourages Type 4s to calm their minds and cultivate mindfulness, creating space for introspection and emotional clarity.

⇒ **Frequency**: Daily for 5-10 minutes.

Open up to Your Dreams

⇒ **Why**: Helps Type 4s explore subconscious feelings and uncover valuable insights through dream reflection.
⇒ **Frequency**: Nightly or as dreams arise.

Open up to Stargazing

⇒ **Why**: Offers Type 4s a sense of wonder and connection to the larger universe, inspiring perspective and emotional balance.
⇒ **Frequency**: Weekly or during clear nights.

Open up to Mantras

⇒ **Why**: Provides a grounding tool for Type 4s to focus their thoughts and redirect emotional energy toward balance and the present moment.
⇒ **Frequency**: Daily for 1-2 minutes.

Type 4: Compatibility With Other Types

Enneagram Type 4, "The Individualist," brings emotional depth, creativity, and a strong desire for authenticity to relationships. Compatibility with other types can vary based on how each type navigates emotions, conflict, and connection.

Type 1 (The Reformer)

•**What Works with Ease**: Type 1s appreciate Type 4's authenticity and passion, while Type 4s respect the integrity and discipline of Type 1s.
•**Potential Friction**: Type 1's critical nature and need for order may clash with Type 4's emotional intensity and desire for creative freedom. Fours may feel judged, while Ones might see Fours as too moody or inconsistent.

Type 2 (The Helper)

•**What Works with Ease**: Type 2s provide emotional support and care, which Type 4s appreciate. Both types value deep, meaningful connections.
•**Potential Friction**: Type 2's desire to help can overwhelm Type 4, who may interpret it as an emotional intrusion. Fours might feel that Twos don't fully understand their emotional complexity, while Twos may feel unappreciated.

Type 3 (The Achiever)

•**What Works with Ease**: Type 3s and 4s share ambition and creativity. Type 3 can help Type 4 channel their emotions into productive goals, while Type 4 can inspire Type 3 to connect with their inner feelings.
•**Potential Friction**: Type 3's focus on success and image can feel shallow to Type 4, who values authenticity. Threes may see Fours as overly sensitive or melancholic, leading to misunderstandings.

Type 4 (The Individualist)

•**What Works with Ease**: Two Type 4s can connect deeply emotionally, sharing creativity and understanding each other's need for individuality.
•**Potential Friction**: Both may struggle with emotional volatility, leading to a relationship marked by intensity, mood swings, and a tendency to withdraw when feelings become overwhelming.

Type 5 (The Investigator)

•**What Works with Ease**: Type 5's intellectual depth can complement Type 4's emotional depth. Both appreciate introspection and solitude, creating a harmonious, quiet dynamic.
•**Potential Friction**: Type 5 may seem emotionally detached to Type 4, while Type 4's emotional intensity can feel draining for Type 5. Misunderstandings may arise if

Type 4 feels unsupported or if Type 5 feels overwhelmed.

Type 6 (The Loyalist)

•**What Works with Ease**: Type 6's loyalty and support help create stability for Type 4, while Type 4's emotional insight can help Type 6 feel understood and reassured.
•**Potential Friction**: Type 6's anxiety and desire for security can feel stifling to Type 4, who seeks freedom and emotional expression. Type 6 may see Type 4 as unpredictable, while Type 4 might feel pressured by Type 6's constant need for reassurance.

Type 7 (The Enthusiast)

•**What Works with Ease**: Type 7 brings energy and excitement into the relationship, encouraging Type 4 to embrace more joy and adventure. Type 4, in turn, can help Type 7 access deeper emotions.
•**Potential Friction**: Type 7's optimism and avoidance of negative emotions can frustrate Type 4, who wants to explore the full range of emotions. Type 7 may find Type 4's emotional intensity draining or overly serious.

Type 8 (The Challenger)

•**What Works with Ease**: Type 8's strength and decisiveness can provide security for Type 4, while Type 4's emotional depth can soften and humanize Type 8. Both share a desire for authenticity.
•**Potential Friction**: Type 8's directness and intensity may overwhelm Type 4, while Type 4's emotional volatility may frustrate Type 8. Type 8's tough exterior can clash with Type 4's sensitivity.

Type 9 (The Peacemaker)

•**What Works with Ease**: Type 9's calm and accepting nature creates a safe space for Type 4 to express their emotions without fear of judgment. Type 4's passion and creativity can inspire Type 9 to deeply engage in life.

•**Potential Friction**: Type 9's tendency to avoid conflict can frustrate Type 4, who seeks emotional engagement and resolution. Type 4 may interpret Type 9's calmness as indifference, while Type 9 may feel overwhelmed by Type 4's emotional intensity.

Depending on the type they interact with, Type 4's emotional intensity and desire for authenticity can create beautiful connections and moments of friction. Relationships thrive when Type 4s feel seen and understood, but challenges arise when their emotional needs are met with avoidance or detachment. Understanding these dynamics can help Type 4s navigate their relationships more smoothly, fostering deeper connections while mitigating friction.

Type 4: Compatibility Table

Pair	Compatibility Dynamics	Strengths Together	Challenges Together	Tips for Improving Flow
4 and 1	Type 1's sense of structure balances Type 4's emotional depth.	Strong shared focus on values and ideals; 1 brings discipline, 4 brings creativity.	1's rigidity may clash with 4's emotional fluidity; 4 may view 1 as overly critical.	Learning to appreciate each other's contrasting styles can foster personal growth and collaboration.
4 and 2	Type 2's nurturing complements Type 4's craving for emotional connection.	Deep emotional connection; 2 provides support, while 4 inspires authenticity.	2 may feel neglected if 4 withdraws into their emotions; 4 may see 2 as overly accommodating.	Balancing giving and receiving in the relationship can create mutual trust and emotional growth.
4 and 3	Type 3's ambition motivates Type 4 to bring their creative visions to life.	High potential for creativity and achievement; 3's drive complements 4's inspiration.	4 may see 3 as too focused on image, while 3 may find 4 overly introspective.	Encouraging vulnerability in 3 and grounding in 4 can strengthen their connection.

Pair	Compatibility Dynamics	Strengths Together	Challenges Together	Growth Opportunities
4 and 4	A shared love for authenticity and emotional depth can lead to a profound connection.	Deep understanding of each other's inner world; mutual appreciation for individuality and creativity.	Both may get stuck in emotional spirals, leading to withdrawal or heightened conflict.	Encouraging practical actions and balancing emotional intensity can stabilize the relationship.
4 and 5	Type 5's analytical nature complements Type 4's emotional insight.	Strong intellectual and emotional bond; 5's detachment balances 4's intensity.	5 may find 4's emotional needs overwhelming, while 4 may see 5 as emotionally distant.	Developing mutual understanding of emotional and intellectual needs fosters a stronger partnership.
4 and 6	Type 6's loyalty reassures Type 4's need for security in relationships.	Mutual devotion and shared sense of support; 6 provides stability, while 4 offers emotional insight.	6 may find 4 unpredictable, while 4 may view 6 as overly anxious or hesitant.	Learning to embrace uncertainty together can help both feel more secure and connected.

Pair	Compatibility Dynamics	Strengths Together	Challenges Together	Growth Opportunities
4 and 7	Type 7's adventurous spirit balances Type 4's introspective nature.	7 brings lightness and spontaneity, while 4 offers emotional depth and reflection.	4 may feel dismissed by 7's avoidance of emotional pain, while 7 may feel constrained by 4's introspection.	Encouraging each other to balance joy and reflection can create a dynamic and fulfilling connection.
4 and 8	Type 8's assertiveness complements Type 4's introspective and emotional qualities.	8's strength provides protection and stability, while 4's depth fosters vulnerability and connection.	4 may feel overwhelmed by 8's intensity, while 8 may see 4 as overly sensitive or impractical.	Building emotional openness in 8 and resilience in 4 can lead to a well-rounded partnership.
4 and 9	Type 9's calm and accepting nature provides grounding for Type 4's emotional intensity.	Deep harmony and understanding; 9's peace balances 4's emotional highs and lows.	4 may see 9 as too passive or disengaged, while 9 may feel drained by 4's emotional fluctuations.	Encouraging 9 to assert themselves and 4 to stabilize their emotions can foster growth and mutual satisfaction.

THE HEAD TRIAD

HEAD Types (5,6,7): Fear of Not Knowing

These types are primarily concerned with security and managing uncertainty, relating to fear and anxiety about the unknown. The core fear is being incapable of coping with life's unpredictability.

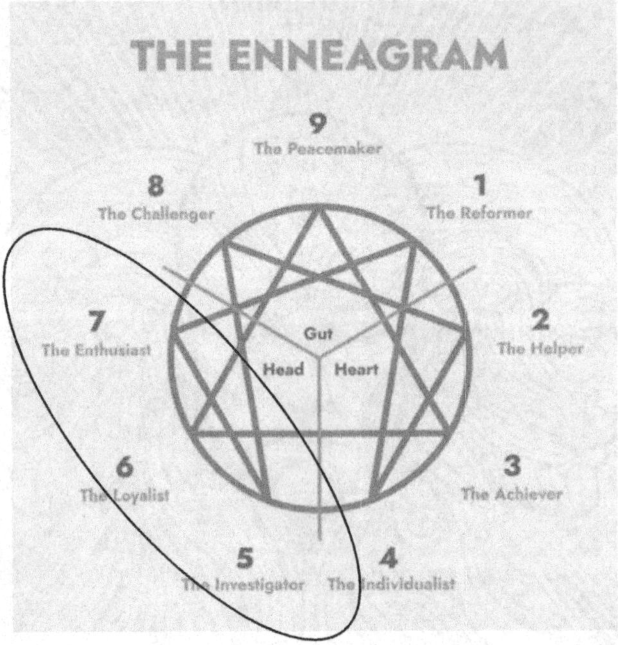

- **Type 5 (The Investigator)**: Their core fear is being helpless or overwhelmed, which drives them to seek knowledge and self-sufficiency. They detach emotionally to focus on gaining expertise.

- **Type 6 (The Loyalist)**: Their core fear is being unsupported or not feeling safe, leading them to seek security through loyalty to people, systems, or beliefs. They often anticipate the worst.

- **Type 7 (The Enthusiast)**: Their core fear is being trapped in pain or deprivation, which drives them to pursue excitement, variety, and positive experiences. They often avoid difficult emotions.

Type 5: The Investigator (Intellectual, observant, innovative, private)

Inquisitive and independent, Investigators seek security through knowledge and self-sufficiency. They are highly observant and enjoy understanding complex ideas. Each type develops unique coping mechanisms—patterns of thought, emotion, and behavior formed to address their core fears and navigate life's challenges. For Investigators, this often involves detaching emotionally and conserving their energy. While this supports intellectual depth, it can lead to difficulty forming close connections or stepping into the unknown.

In the following sections, we explore how Type 5 typically operates in key areas of life—at work, home, and in relationships—revealing its characteristic strengths, vulnerabilities, and tendencies.

We also examine this type's consciousness and plasticity challenges, identifying opportunities to enhance self-awareness, adaptability, and emotional intelligence. To support these efforts, we provide tailored *OPEN rituals*—PRIMAL, RATIONAL, and SPIRITUAL practices that nurture openness, balance, and personal growth.

Finally, the section concludes by unpacking compatibility dynamics, offering insights into how Type 5 relates to all other Enneagram types across various professional, familial, or romantic relationships.

The Gifts of Being an Enneagram 5

1. **Intellectual Depth and Curiosity:** Type 5s are known for their insatiable curiosity and love of learning. They are natural thinkers who seek to understand the world deeply and intellectually. This gift allows them to excel in fields that require problem-solving, analysis, and expertise. Their ability to delve into complex subjects makes them incredibly resourceful, often becoming the "go-to" person in their area of knowledge. Type 5s are lifelong learners who thrive on mastering new information, enabling them to bring fresh, innovative ideas to any situation, personal or professional.

2. **Independence and Self-Sufficiency:** Type 5s value their independence and have a strong sense of self-sufficiency. They can work autonomously, often preferring solitude to recharge and think through problems. Their self-reliance allows them to avoid becoming overly dependent on others, which makes them excellent at managing themselves and their projects. In relationships, this independence allows Type 5s to give others space and autonomy, fostering mutual respect. They are comfortable with their own company, and this trait can serve as a grounding force for both them and others.

3. **Insight and Objectivity:** Type 5s can approach situations with objectivity and detachment, allowing them to analyze situations without emotion. This is a tremendous gift in decision-making, as they can sift through information, make logical conclusions, and offer insightful perspectives. In relationships, their ability to provide thoughtful advice without getting emotionally entangled allows them to be supportive partners and friends. Their insight often brings clarity to chaotic or uncertain situations, making them valuable problem-solvers in personal and professional settings.

4. **Innovative Thinking and Creativity:** Though they are often seen as intellectuals, Type 5s are also deeply creative. Their ability to connect seemingly unrelated ideas and draw from their vast knowledge base allows them to think outside the box. They can approach challenges with innovative solutions, often seeing patterns and opportunities others might overlook. Their creativity may manifest in areas like technology, science, art, or philosophy, where they excel at offering original and imaginative contributions.

Type 5 Coping Mechanisms

Type 5s, the Investigators, often cope by withdrawing from external demands and conserving their energy. They seek knowledge and self-sufficiency as a way to feel secure in a world that can seem overwhelming. While this strategy helps them maintain a sense of control, it can also lead to emotional detachment, isolation, and difficulty relying on others.

1. Emotional Detachment

- **Coping Mechanism**: As children, Type 5s learned to detach emotionally to avoid feeling overwhelmed or vulnerable. By withdrawing into their minds, they created a sense of security where they could observe the world from a distance rather than actively engage with it.
- **Adult Challenge**: In adulthood, this detachment can make it difficult to express emotions and form deep, meaningful connections. Type 5s may struggle with intimacy, fearing that emotional closeness will deplete their internal resources or expose them to discomfort.

2. Focus on Knowledge

- **Coping Mechanism**: Type 5s found safety and control through learning and intellectual pursuits. Accumulating knowledge became a way to feel prepared for any situation and avoid the unpredictability of emotional interactions.
- **Adult Challenge**: This can result in over-reliance on research and analysis, sometimes at the expense of action. Type 5s may struggle with decision-making, fearing they don't know enough to move forward. Their focus on knowledge can also lead to a disconnect from their own emotions and physical needs.

3. Conserving Energy

- **Coping Mechanism**: From a young age, Type 5s felt they had limited emotional and physical energy, leading them to withdraw and carefully manage their engagements. They learned to retreat to solitude to recharge.
- **Adult Challenge**: This habit can cause them to hold back in relationships and work, fearing that too much interaction will drain them. They may hesitate to engage deeply with others or avoid situations requiring sustained emotional effort.

4. Avoidance of Dependency

- **Coping Mechanism**: To maintain a sense of control, Type 5s often avoided relying on caregivers or authority figures. They developed an intense need for independence, believing that self-sufficiency was the only way to stay safe.
- **Adult Challenge**: This deep-seated fear of dependence can lead to difficulty asking for help, even when they need it. They may struggle with collaboration or vulnerability, resisting the idea of relying on others for support.

5. Overthinking as a Defense

- **Coping Mechanism**: Type 5s developed a habit of overanalyzing situations to feel prepared and avoid emotional discomfort. Thinking became a way to feel safe, serving as a buffer between them and the unpredictability of life.
- **Adult Challenge**: This tendency can lead to indecision, procrastination, and disconnection from emotions. Type 5s may struggle to act in the moment, feeling the need to gather more information before making even small choices.

6. Isolation as Protection

- **Coping Mechanism**: Many Type 5s created a private inner world where they felt safe from criticism, intrusion, or excessive demands. They learned to retreat into their thoughts rather than seek external validation.
- **Adult Challenge**: This can result in loneliness and difficulty sharing their inner world with others. Type 5s may find themselves isolated, struggling to bridge the gap between intellectual pursuits and emotional engagement.

By recognizing these coping mechanisms, Type 5s can work toward balancing their need for independence with healthy connection. Learning to trust their own emotions, engage in the present, and open up to support from others can help them cultivate a more fulfilling and engaged life.

Type 5 at Work

Type 5s bring a wealth of knowledge, creativity, and

analytical prowess to the workplace. They thrive in environments that value intellectual depth and autonomy, often excelling in roles that require problem-solving and innovation. However, their preference for independence and emotional detachment can sometimes create collaboration and communication challenges. Understanding their strengths and growth areas can help Type 5s reach their full potential in professional settings.

Strengths

⇒ **Analytical Thinking:** Type 5s excel in processing complex information, making them exceptional problem-solvers and strategists.

⇒ **Focus and Depth:** They dive deeply into their work, mastering their areas of interest and producing high-quality results.

⇒ **Innovative and Creative:** Their intellectual curiosity often leads to unique insights and innovative solutions.

⇒ **Independent Work Style:** Type 5s thrive in roles that allow autonomy, where they can self-manage and focus on tasks without frequent oversight.

⇒ **Objective Decision-Making:** They value logic and facts over emotions, helping them make well-reasoned decisions.

Challenges

⇒ **Emotional Detachment:** Their preference for independence and objectivity can make them seem aloof or unapproachable to colleagues.

⇒ **Reluctance to Share Knowledge:** Type 5s may hoard information, fearing they lack sufficient expertise to contribute fully or being overly protective of their ideas.

⇒ **Avoidance of Collaboration:** They often prefer solitary work and may resist team efforts, especially if the dynamics feel emotionally demanding.

⇒ **Overthinking and Indecision:** Their tendency to analyze extensively can lead to delays in decision-making and missed opportunities.

⇒ **Energy Conservation:** Type 5s can limit their involvement in work tasks or relationships to avoid feeling drained, which might be perceived as disengagement.

Type 5 at Home

At home, Type 5s are thoughtful, reflective, and dedicated to creating a peaceful environment. They value their personal space and enjoy intellectual pursuits, often bringing a sense of calm and curiosity to family life. However, their tendency to withdraw emotionally or conserve energy can make connecting deeply with loved ones challenging. Balancing their need for independence with emotional engagement can help them foster more fulfilling relationships at home.

Strengths

⇒ **Thoughtful and Insightful:** Type 5s bring depth and curiosity to family life, often offering unique perspectives and creative ideas.

⇒ **Calm and Composed:** They maintain emotional steadiness, providing stability during conflicts or stressful situations.

⇒ **Support for Independence:** Type 5s respect boundaries and encourage loved ones to explore their interests and autonomy.

⇒ **Problem-Solving:** They analyze family challenges and offer logical, well-thought-out solutions.

⇒ **Low-Demand Presence:** Type 5s are not overly demanding attention, creating space for others to express themselves freely.

Challenges

⇒ **Emotional Withdrawal:** Their tendency to retreat into their inner world can leave family members feeling disconnected or unsupported.

⇒ **Difficulty Expressing Emotions:** Type 5s may struggle to share their feelings, creating barriers to emotional intimacy with loved ones.
⇒ **Over-prioritizing Personal Space:** Their need for solitude can sometimes come across as neglecting family responsibilities or relationships.
⇒ **Conserving Energy:** They may hesitate to engage deeply in family activities, fearing emotional or physical exhaustion.
⇒ **Over-intellectualizing:** Type 5s may approach emotional or relational issues logically, missing opportunities for empathetic connection.

Type 5 in Relationship

Type 5s are loyal, introspective, and supportive in relationships, often bringing depth and intellectual curiosity to their partnerships. They value independence and mutual respect, thriving in relationships that allow personal space and intellectual connection. However, their tendency to withdraw emotionally and avoid vulnerability can make it challenging for partners to feel deeply connected. Cultivating emotional openness and balancing solitude with intimacy are key to fostering fulfilling relationships.

Strengths

⇒ **Loyal and Committed:** Type 5s are dedicated to their partners and strive to be dependable and trustworthy in their relationships.
⇒ **Intellectual Connection:** They bring thoughtfulness and stimulating conversation, enriching their relationships with insight and curiosity.
⇒ **Respect for Independence:** Type 5s encourage autonomy in their partners, avoiding clinginess or over-dependence.
⇒ **Calm and Rational:** They approach conflicts and challenges with a composed and analytical mindset, often providing balanced solutions.
⇒ **Low Maintenance:** Type 5s require little emotional drama or attention, preferring harmonious, straightforward

connections.

Challenges

⇒ **Emotional Withdrawal:** Their tendency to retreat during conflicts or emotional situations can leave partners feeling unsupported or neglected.

⇒ **Difficulty with Vulnerability:** Type 5s often avoid sharing their inner emotional world, making it hard for their partners to understand them truly.

⇒ **Limited Emotional Expression:** They may focus more on logic than feelings, leading to a perceived lack of warmth or empathy in the relationship.

⇒ **Overemphasis on Personal Space:** Type 5s' need for solitude can sometimes feel like emotional distance, straining intimacy.

⇒ **Fear of Dependence:** They may hesitate to rely on their partner or fear their partner becoming too reliant, creating barriers to interdependence.

By embracing emotional vulnerability and actively engaging in relationships, Type 5s can deepen their emotional connections and build stronger, more fulfilling partnerships.

Type 5 Consciousness Challenges

For **Type 5 (The Investigator)**, the challenges related to consciousness and neuroplasticity are centered around their tendencies toward emotional detachment, intellectualization, and a desire to conserve their resources (energy, time, knowledge). These patterns influence their capacity to expand their consciousness and rewire their brains for greater emotional openness and connection with others.

1. Emotional Detachment and Isolation

Type 5s often rely on emotional detachment to protect themselves from feeling overwhelmed by the demands of others or by their environment. This detachment narrows their **consciousness** by preventing them from fully engaging with their emotions and the emotional experiences of others. While their intellectual depth helps them understand the world, it also creates a barrier to expanding their awareness into the emotional and relational realms. Type 5s may struggle to access higher states of **consciousness** like compassion and empathy because they are more comfortable analyzing emotions than feeling them.

2. Tendency to Intellectualize Experience

One of the key coping mechanisms for Type 5 is to intellectualize their experiences, turning everything into something to be studied, understood, or analyzed. This can limit their ability to fully engage with the present moment as they retreat into their minds to make sense of everything. Consciousness expansion requires a balance between thought and presence. Still, Type 5s often find themselves disconnected from their bodies and emotions, viewing life from a mental perspective rather than an embodied one. This intellectual distancing can prevent them from accessing states of consciousness that require emotional vulnerability or deep relational awareness.

3. Conserving Energy and Withdrawing

Type 5s are often preoccupied with the idea that their energy is limited and they need to conserve their resources. This belief makes them withdraw from social interactions and emotional experiences, which can keep their consciousness focused on self-protection rather than expansion. They may avoid situations that require emotional investment or vulnerability because they fear being depleted. This focus on conserving energy restricts their ability to open up, experience emotional connections, and trust the flow of life, keeping their consciousness in a state of self-imposed limitation.

Type 5: Plasticity Challenges

Neuroplasticity refers to the brain's ability to reorganize itself by forming new neural connections throughout life. It allows people to change habits, beliefs, and emotional responses through new experiences and learning.

1. Rewiring Emotional Detachment into Engagement

One key challenge for Type 5 is rewiring their tendency toward emotional detachment. Their brain has developed strong neural pathways that reinforce withdrawal and isolation, making it difficult for them to engage with emotions directly and presently. To change this, Type 5s must consciously practice emotional openness and vulnerability, which can feel uncomfortable for them. They need to cultivate new neural circuits that support emotional engagement and connection with others, allowing them to break free from their pattern of avoidance.

2. Overcoming the Tendency to Intellectualize

Type 5s' habit of intellectualizing their feelings and experiences can become deeply entrenched, as it serves as a protective mechanism. This pattern creates well-worn neural pathways that favor analysis over emotional experience. For Type 5 to engage in neuroplasticity, they must shift from intellectualizing emotions to feeling them directly and consciously. This means developing practices encouraging embodiment—being present in the body and emotions rather than retreating into the mind. Mindfulness, breathwork, or body-focused practices like yoga can help Type 5s create new neural connections that bring them out of their heads and into the present moment.

4. Breaking the Cycle of Withdrawal

Type 5s often withdraw from situations that require emotional or relational energy because they fear being overwhelmed or depleted. This withdrawal reinforces neural pathways that prioritize solitude and self-protection over engagement. For neuroplasticity to take effect, Type 5s must practice stepping into situations that challenge their comfort zone, particularly in relationships and social interactions. By gradually exposing themselves to relational experiences and learning to trust that they can manage these interactions without becoming depleted, Type 5s can create new neural pathways that support connection and emotional resilience.

5. Letting Go of the Scarcity Mindset

Type 5s often operate from a scarcity mindset, believing their time, energy, or resources are limited. This mindset creates neural patterns that reinforce withdrawal and conservation. To rewire these patterns, Type 5s must practice abundance thinking, which involves challenging the belief that they will be drained if they give of themselves emotionally or intellectually. Through small acts of generosity—whether in sharing time, knowledge, or emotional energy—Type 5s can begin to develop neural connections that support a more open and generous mindset, gradually reducing their fear of depletion.

For Type 5, the primary challenges about consciousness and neuroplasticity stem from their habitual emotional detachment, intellectualization, and fear of being overwhelmed. These patterns narrow their consciousness by keeping them focused on self-protection rather than emotional engagement, and they reinforce neural pathways that prioritize withdrawal over connection. Type 5s must practice emotional vulnerability, embodiment, and abundance thinking to expand their consciousness and engage neuroplasticity. By doing so, they can develop new neural pathways that foster deeper relationships, emotional resilience, and a more expansive sense of self.

Type 5: Summary of Challenges

Challenges of Type 5	Impact on Consciousness	Impact on Neuroplasticity
Emotional Detachment	Limits emotional engagement and prevents deeper connections with others.	Reinforces avoidance of emotions; difficult to form new pathways for vulnerability and openness.
Tendency to Intellectualize	Disconnects from present-moment experiences by over-analyzing rather than feeling.	Strengthens analytical pathways, making it harder to engage with emotions directly.
Conserving Energy and Withdrawing	Narrows consciousness to self-protection, limiting emotional and relational growth.	Reinforces patterns of withdrawal; makes it challenging to build neural circuits for engagement.
Fear of Being Overwhelmed	Keeps consciousness focused on scarcity, limiting openness to life's experiences.	Strengthens fear-based pathways; prevents formation of circuits that support emotional resilience.
Scarcity Mindset	Reduces ability to embrace abundance and openness, focusing instead on limits.	Hardens existing neural circuits of self-preservation; hinders creation of pathways that support generosity and trust.

Type 5: PRIMAL Rituals Recommended

For Type 5s, PRIMAL rituals reconnect them with their physical body and emotional world. These rituals encourage them to soften their detachment and embrace vulnerability by focusing on sensory experiences and presence.

Open up to Your Breath

⇒ **Why:** Breathing exercises calm the nervous system and reduce overstimulation.
⇒ **Frequency:** 5–10 minutes daily.

Open up to Your Sleep

⇒ **Why:** Prioritizing restorative sleep aids emotional resilience and intellectual clarity.
⇒ **Frequency:** Nightly, ensuring 7–9 hours of uninterrupted sleep.

Open up to Nature

⇒ **Why:** Spending time outdoors recharges energy and provides a perspective beyond the intellectual focus.
⇒ **Frequency:** Weekly walks or visits to natural settings.

Type 5: RATIONAL Rituals Recommended

RATIONAL rituals guide Type 5s toward openness and trust. These practices reduce their tendency to withdraw and invite them to share their thoughts and feelings, fostering deeper emotional engagement.

Open up to Learning

⇒ **Why:** Learning new skills or exploring new areas of interest stimulates intellectual engagement and promotes cognitive flexibility.

⇒ **Frequency:** Weekly (1–2 hours).

Open up to Cooking

⇒ **Why:** Cooking provides a creative and sensory outlet, helping Type 5s stay grounded and engaged in the present moment.
⇒ **Frequency:** Weekly (1–2 times).

Open up to Laughter

⇒ **Why:** Humor and laughter enhance emotional connection, reduce stress, and help Type 5s balance their cerebral tendencies with lightheartedness.
⇒ **Frequency:** Weekly, during social interactions.

Type 5: SPIRITUAL Rituals Recommended

SPIRITUAL rituals for Type 5s emphasize surrendering the need for certainty and control. Through contemplative activities, they can explore interconnectedness and expand beyond intellectual boundaries.

Open up to Meditation

⇒ **Why:** Meditation quiets the mind, fosters inner calm, and enhances emotional clarity, reducing the tendency toward detachment.
⇒ **Frequency:** Daily (10–15 minutes).

Open up to Prayer

⇒ **Why:** Engaging in prayer or spiritual reflection helps Type 5s surrender control and trust in a force beyond their intellect.
⇒ **Frequency:** Daily (5–10 minutes).

Open up to Rituals

- ⇒ **Why:** Personal rituals (e.g., lighting candles or journaling) provide mindfulness and structure, encouraging presence and emotional openness.
- ⇒ **Frequency:** Weekly.

Type 5: Compatibility With Other Types

Type 1 (The Reformer)

•**What Works**: Both types value independence, self-discipline, and precision. Type 5's intellectual approach complements Type 1's principled nature; together, they can have deep, meaningful discussions. Type 1 appreciates Type 5's thoughtful and detailed thinking.
•**Potential Friction**: Type 1's tendency toward perfectionism may feel restrictive to Type 5, who prefers autonomy and freedom to explore ideas. Type 5's emotional detachment may frustrate Type 1, who values clear standards and emotional responsibility.

Type 2 (The Helper)

•**What Works**: Type 2's warmth and generosity can draw Type 5 out of their intellectual shell, encouraging more emotional engagement. Type 5 appreciates Type 2's caring nature, while Type 2 admires Type 5's intelligence and thoughtful perspective.
•**Potential Friction**: Type 5 may find Type 2's emotional needs overwhelming, as they tend to guard their energy. If Type 5 withdraws too often, Type 2 may feel neglected or unloved, leading to frustration or feelings of rejection.

Type 3 (The Achiever)

•**What Works**: Both are driven by excellence, though differently. Type 5 brings deep analysis and thoughtful decision-making to the relationship, while Type 3 provides energy and momentum to get things done. Together, they can accomplish significant goals.

•**Potential Friction**: Type 5's need for solitude and careful planning may clash with Type 3's fast-paced, success-oriented approach. Type 3's focus on external validation and achievement may feel shallow to Type 5, who values depth and authenticity.

Type 4 (The Individualist)

•**What Works**: Both types are introspective and enjoy deep emotional and intellectual discussions. Type 5's intellectual depth and Type 4's emotional intensity can create a strong, creative partnership in which each respects the other's inner world.
•**Potential Friction**: Type 4's emotional volatility can be overwhelming for the more detached Type 5. Similarly, Type 5's tendency to withdraw emotionally may leave Type 4 feeling abandoned or misunderstood, which can lead to conflict.

Type 5 (The Investigator)

•**What Works**: Two Type 5s can form a harmonious relationship based on mutual respect for boundaries, intellectual exploration, and independence. They can engage in meaningful conversations and enjoy plenty of space without feeling neglected.
•**Potential Friction**: Both may struggle to meet each other's emotional needs, as neither is likely to initiate deep emotional connection. Their shared tendency to withdraw could lead to isolation, where both avoid confronting emotional issues.

Type 6 (The Loyalist)

•**What Works**: Type 6's loyalty and Type 5's self-sufficiency can create a stable, supportive relationship. Type 6 appreciates Type 5's analytical mind, while Type 5 values Type 6's commitment and reliability.
•**Potential Friction**: Type 6's **anxiety** and need for reassurance may frustrate Type 5, who values emotional independence. Type 6 may feel insecure if Type 5 withdraws emotionally, leading to trust issues.

Type 7 (The Enthusiast)

•**What Works**: Type 7's energetic, optimistic nature can help pull Type 5 out of their introspection and encourage them to enjoy life more fully. Type 5's thoughtful approach can help ground Type 7's spontaneity.
•**Potential Friction**: Type 7's constant need for **stimulation** and adventure can exhaust Type 5, who prefers quiet, contemplative environments. Type 7 may find Type 5 too detached and serious, while Type 5 could see Type 7 as scattered and unfocused.

Type 8 (The Challenger)

•**What Works**: Type 8's strength and decisiveness can help Type 5 feel protected and empowered to explore their intellectual pursuits. Type 5's ability to strategize and think critically complements Type 8's drive for action.
•**Potential Friction**: Type 8's **dominant personality** may overwhelm Type 5, who prefers to avoid conflict and maintain autonomy. Type 5 may withdraw if Type 8 becomes too confrontational or controlling, leading to tension.

Type 9 (The Peacemaker)

•**What Works**: Both types value **peace and autonomy**. Type 9's calm, easygoing nature allows Type 5 to feel comfortable and unpressured, while Type 5's self-sufficiency aligns with Type 9's desire for harmony.
•**Potential Friction:** Type 9's tendency to avoid conflict may frustrate Type 5, especially if important issues go unresolved. Type 5's withdrawal and emotional detachment can also leave Type 9 feeling disconnected or neglected.

Type 5: Compatibility Table

Pair	Compatibility Dynamics	Strengths Together	Challenges Together	Tips for Improving Flow
5 and 1	A shared love for logic and structure fosters mutual respect, though emotional detachment may hinder connection.	Rational problem-solving, shared value for independence.	Difficulty addressing emotional needs; rigid expectations.	Learning to balance emotional connection with intellectual alignment.
5 and 2	Type 5's independence complements Type 2's emotional generosity, but differing energy levels can clash.	Deeply loyal partnership, blending emotional warmth with intellectual depth.	Type 5 may withdraw from Type 2's emotional intensity; Type 2 may feel neglected.	Communicating needs openly and balancing independence with emotional availability.
5 and 3	Intellectual pursuits of Type 5 align with Type 3's ambitious drive, but priorities may diverge.	High levels of productivity, combining strategic thinking with action-oriented goals.	Type 5's slower pace may frustrate Type 3; Type 3's need for recognition may feel excessive to Type 5.	Aligning goals and respecting each other's unique approaches to success.

Pair	Compatibility Dynamics	Strengths Together	Challenges Together	Tips for Improving Flow
5 and 4	A natural pairing as Type 5's analytical nature complements Type 4's emotional depth, though misunderstandings may arise.	Creativity and innovation; Type 5 provides structure, while Type 4 brings emotional insight.	Type 5 may struggle with Type 4's emotional intensity; Type 4 may feel Type 5 is too distant.	Building mutual understanding by balancing intellect with emotional depth.
5 and 5	Shared intellectual curiosity creates harmony, but both may avoid emotional connection.	Deep understanding and respect for personal space; shared passion for knowledge.	Emotional detachment , difficulty expressing feelings or addressing relational needs.	Encouraging vulnerability and balancing intellectual connection with emotional intimacy.
5 and 6	A strong mental connection is rooted in analyzing and preparing, but trust issues may create distance.	Complementary problem-solving skills; mutual loyalty.	Type 6's anxiety may overwhelm Type 5; Type 5's withdrawal may frustrate Type 6.	Strengthening trust by addressing fears and committing to open communication.
5 and 7	Contrasting energies: Type 5's focus meets Type 7's spontaneity, creating balance or conflict depending on circumstances.	Innovation and exploration; Type 5 grounds Type 7, while Type 7 inspires Type 5 to take risks.	Type 7 may view Type 5 as overly cautious; Type 5 may see Type 7 as too scattered.	Learning to appreciate and integrate each other's contrasting approaches to life.

Pair	Compatibility Dynamics	Strengths Together	Challenges Together	Tips for Improving Flow
5 and 8	Type 5's intellectual independence balances Type 8's assertiveness, but power struggles may arise.	Strategic collaboration, combining Type 5's analytical mind with Type 8's decisiveness.	Type 8 may dominate, causing Type 5 to withdraw; Type 5's detachment may frustrate Type 8.	Cultivating mutual respect by valuing intellect alongside assertiveness.
5 and 9	A peaceful partnership where Type 5's focus complements Type 9's calm demeanor, but inertia can slow progress.	Harmonious and low-conflict dynamic; mutual respect for independence and personal space.	Lack of initiative; avoidance of difficult emotional conversations.	Encouraging each other to engage more actively in the relationship and address challenges proactively.

Type 6: The Loyalist (Devoted, cautious, reliable, apprehensive)

Faithful and vigilant, Investigators seek security and stability, often finding reassurance through trust in relationships or systems. They are practical and responsible. Each type develops unique coping mechanisms—patterns of thought, emotion, and behavior formed to address their core fears and navigate life's challenges. For Type 6, this often involves anticipating potential problems and relying on external support. While this fosters preparedness and loyalty, it can also lead to anxiety or overdependence.

In the following sections, we explore how Type 6 typically operates in key areas of life—at work, home, and in relationships—revealing their characteristic strengths, vulnerabilities, and tendencies. We also examine this type's consciousness and plasticity challenges, identifying opportunities to enhance self-awareness, adaptability, and emotional intelligence. To support these efforts, we provide tailored *OPEN rituals*—PRIMAL, RATIONAL, and SPIRITUAL practices that nurture openness, balance, and personal growth.

Finally, the section concludes by unpacking compatibility dynamics, offering insights into how Type 6 relates to all other Enneagram types across various professional, familial, or romantic relationships.

The Gifts of Being an Enneagram 6

1. **Loyalty and Dependability**: Type 6s are known for their deep commitment and loyalty to the people and causes they care about. Once they feel secure in their relationships or workplace, they are reliable and trustworthy and will go to great lengths to protect and support others. This makes them highly valued as friends, partners, and employees who can be counted on during challenging times.

2. **Preparedness and Problem-Solving**: Type 6s naturally can anticipate potential risks and challenges. Their attention to detail and strategic thinking enable them to plan for the future,

preparing for various outcomes. This foresight makes them excellent planners and problem-solvers, able to foresee difficulties and create solutions before problems arise.

3. **Collaboration and Teamwork:** Type 6s are naturally inclined to work within groups and value teamwork and community. They thrive in environments with mutual trust and support and excel at fostering team cooperation and cohesion. Their collaborative nature means they are open to feedback and willing to compromise for the greater good.

4. **Protectiveness and Advocacy:** With their strong sense of duty and concern for others, Type 6s often take on a protective role, advocating for those they feel are vulnerable or need support. They are excellent at standing up for their loved ones or defending their beliefs, often putting others' needs ahead of their own.

Type 6 Coping Mechanisms

Type 6s, the Loyalists, often cope by anticipating potential threats and seeking security through external support. They develop patterns of worry, skepticism, and reliance on trusted figures to navigate their fear of uncertainty. These coping mechanisms provide them with a sense of preparedness but can also reinforce their anxieties and dependence if not consciously managed.

1. Anxiety and Overthinking

- **Coping Mechanism**: As children, Type 6s learned to be hypervigilant about their surroundings, constantly scanning for potential threats or problems. This heightened awareness made them cautious and analytical but also contributed to chronic anxiety. Overthinking became their way of feeling prepared and in control.
- **Adult Challenge**: In adulthood, this tendency often manifests as cycles of worry and indecision. Type 6s may struggle with excessive doubt, second-guessing their choices, and fearing worst-case scenarios. Their reliance on

others for reassurance can make it difficult for them to trust their instincts and make independent decisions.

2. Difficulty Trusting Themselves and Others

- **Coping Mechanism**: Growing up in unpredictable environments, Type 6s often sought security by relying on authority figures or trusted individuals for guidance. This dependency provided them with stability and a sense of protection.
- **Adult Challenge**: As adults, Type 6s may continue to look to others for validation, struggling to trust their own judgment. This pattern can lead to dependency in relationships or workplaces, making it hard for them to assert themselves confidently. Conversely, they may also become skeptical or distrustful, fearing betrayal or incompetence from those they rely on.

3. Fear of Betrayal and Abandonment

- **Coping Mechanism**: Many Type 6s grew up with an underlying fear of being abandoned or deceived. To protect themselves, they developed a habit of questioning others' motives and remaining on guard for signs of betrayal.
- **Adult Challenge**: This fear can lead to suspicion, defensiveness, or excessive testing of loyalty in relationships. While their vigilance helps them avoid genuine threats, it can also push people away or create unnecessary tension. Some Type 6s respond by becoming overly loyal, holding onto relationships—even unhealthy ones—for fear of instability.

4. Complex Relationship with Authority

- **Coping Mechanism**: In childhood, Type 6s often had a conflicting relationship with authority figures, simultaneously seeking their protection while fearing their failure or unpredictability. This duality led them to either strictly adhere to rules or resist them in moments of doubt.
- **Adult Challenge**: As adults, this dynamic can cause them to either overly comply with authority or rebel against it,

depending on their level of trust. They may struggle in hierarchical structures, fearing both control and abandonment by leadership. Finding a balanced approach to authority—where they can respect guidance while maintaining autonomy—is key to their personal growth.

By recognizing these coping mechanisms, Type 6s can begin to build inner trust and resilience. Learning to balance caution with confidence and external support with self-reliance helps them navigate life with greater ease and security.

Type 6 at Work

Type 6s, known for their loyalty and vigilance, bring a sense of dependability and foresight to the workplace. They are naturally inclined to anticipate challenges and work toward creating a stable, secure environment. However, their focus on safety and certainty can sometimes lead to overthinking or hesitancy. Below is a detailed look at their **strengths** and **challenges** in professional settings:

Strengths

⇒ **Reliability and Teamwork**: Type 6s are highly dedicated and dependable, often becoming the "glue" that holds teams together.
⇒ **Planning and Risk Management**: They excel at foreseeing potential issues, making them excellent planners and troubleshooters.
⇒ **Thorough and Organized**: Their detail-oriented nature ensures tasks are completed methodically and efficiently.
⇒ **Protective and Advocative**: They stand up for others and prioritize creating a safe, stable workplace environment.

Challenges

⇒ **Over-caution and Anxiety**: Their focus on security can lead to overthinking and hesitancy in decision-making.
⇒ **Dependence on Guidance**: When leadership is unclear, they may rely too heavily on others for reassurance or guidance.

⇒ **Difficulty with Authority**: They may struggle to trust or feel confident in authority figures, leading to either over-reliance or skepticism.

⇒ **Self-Doubt**: Their tendency to second-guess themselves can slow progress and undermine their confidence.

Type 6 at Home

Type 6s, known for their dedication and vigilance, strive to create a safe, stable, and supportive home environment. Their protective instincts ensure their loved ones feel secure, though their desire for certainty can sometimes lead to stress or over-caution. Below are their strengths and challenges at home:

Strengths

⇒ **Loyal and Protective**: Type 6s are deeply committed to their family and work tirelessly to ensure the safety and well-being of their loved ones.

⇒ **Planning and Organization**: They excel at organizing and planning for the family's future, ensuring consistency and stability.

⇒ **Reliable and Trustworthy**: Type 6s are known for their dependability and offer emotional and practical support in times of need.

⇒ **Comforting Routines**: Their ability to establish and maintain routines brings comfort and reassurance to their household

Challenges

⇒ **Over-Worrying**: Their focus on potential risks can lead to unnecessary stress for themselves and others in the household.

⇒ **Controlling Behavior**: In their effort to ensure safety, they may become overly controlling, limiting spontaneity or flexibility in family dynamics.

⇒ **Excessive Reassurance-Seeking**: Type 6s may look to family members for constant validation, which can strain relationships over time.

⇒ **Resistance to Change**: Their desire for security may cause them to resist beneficial changes, even when they could improve the household dynamic.

Type 6 in Relationships

Type 6s are devoted and protective partners, deeply committed to creating stable and supportive relationships. Their loyalty and focus on security make them dependable companions. However, their anxiety and need for reassurance can sometimes create challenges. Below is a breakdown of their strengths and challenges in relationships:

Strengths

⇒ **Loyal and Committed**: Type 6s deeply value trust and emotional security, making them reliable and devoted partners.

⇒ **Protective and Supportive**: They are nurturing and prioritizing the safety and well-being of their loved ones, offering steadfast emotional support during challenging times.

⇒ **Preparedness**: Their ability to anticipate problems helps ensure relationship stability, as they proactively address potential issues before they arise.

Challenges

⇒ **Anxiety and Reassurance-Seeking**: Their tendency to overthink and fear abandonment can lead to seeking constant reassurance, which may strain the relationship.

⇒ **Defensiveness or Suspicion**: Driven by worst-case scenarios, they might become overly cautious or mistrustful, even when there's no real threat.

⇒ **Fear of Betrayal**: Deep-seated fears of betrayal or insecurity can result in tension, as they may test their partner's loyalty or unintentionally create distance.

Type 6 Consciousness Challenges

For **Type 6 (The Loyalist)**, the specific challenges related to consciousness and neuroplasticity center around their habitual tendencies toward fear, doubt, and reliance on external security sources. These deeply ingrained patterns impact their ability to shift into higher states of awareness and rewire their brains for healthier, more empowered thinking.

1. Fear-Based Thinking and Vigilance

Type 6s operate from a fear-driven mindset, often focused on worst-case scenarios. This can create a narrowed consciousness, where their mental energy is consumed by what might go wrong. Their hyper-vigilance to potential threats makes it difficult for them to stay present and trust in the flow of life. This constant scanning for danger limits their ability to access higher states of awareness, such as mindfulness or self-trust, which are essential for expanding consciousness beyond survival thinking.

2. Reliance on External Authority and Reassurance

A core issue for Type 6 is their tendency to outsource decision-making to others, whether it be authority figures, loved ones, or trusted friends. This creates a block in consciousness development, as they do not fully trust their own inner guidance. Because of their habit of seeking reassurance, they struggle to connect deeply with their intuition and inner wisdom. This reliance on external validation prevents them from reaching a more autonomous, awakened state of consciousness where self-trust and inner knowing are central.

3. Emotional Reactivity and Fear of Betrayal

Type 6s often live with an undercurrent of anxiety and emotional reactivity, particularly with perceived threats or betrayals. This emotional volatility can cloud their consciousness, making it harder for them to respond calmly and clearly in stressful situations. The constant cycling of anxiety keeps them locked in a low-frequency emotional state that impedes their ability to shift into higher awareness, where they can recognize that their fears are often unfounded.

Type 6: Plasticity Challenges

Neuroplasticity refers to the brain's ability to reorganize itself by forming new neural connections throughout life. It allows people to change habits, beliefs, and emotional responses through new experiences and learning.

1. Overcoming Conditioned Fear Responses

Neuroplasticity allows the brain to rewire itself through new experiences and learning. For Type 6s, one of the biggest challenges is unlearning their conditioned fear responses. These habitual reactions to perceived danger or insecurity are deeply ingrained in their brains' neural pathways. To rewire these pathways, Type 6s must practice stepping into trust—trusting themselves, others, and the world around them. This is a slow process, as the brain must learn to replace fear-based patterns with more resilient and empowering neural circuits.

2. Rewiring the Need for External Validation

A major obstacle for Type 6s is their dependence on external reassurance. This results from repeated reliance on external sources for safety and validation in their formative years. To engage in neuroplasticity, Type 6s must practice self-reliance and inner trust. This involves training the brain to stop seeking answers or approval from others and instead rely on their judgment. By consciously shifting their responses to uncertainty, they can gradually rewire their brains to operate from a place of self-confidence rather than fear.

3. Breaking the Cycle of Anxiety and Worry

The brain of Type 6 is often wired for anxiety, constantly generating thoughts about potential dangers or risks. This mental pattern reinforces itself over time, creating a well-worn path of worry in the brain. For Type 6s to change this pattern, they must engage in practices that calm the nervous system and interrupt the cycle of anxiety. Mindfulness, meditation, and deep breathing can help to create new neural connections that foster calmness and clarity. These practices promote neuroplasticity by redirecting the brain's focus from fear-based thoughts to more balanced and present-moment awareness.

Type 6 faces specific challenges in consciousness and plasticity due to their deeply ingrained fear patterns, need for external validation, and chronic anxiety. Developing self-trust and practicing mindfulness are key areas where Type 6s can work to expand their consciousness and rewire their brains. Over time, by consciously shifting their focus from fear to trust, Type 6s can create new, more empowering neural pathways that support greater inner peace, resilience, and self-reliance.

Type 6: Summary of Challenges

Challenge	Impact on Consciousness	Impact on Plasticity
Fear-Based Thinking	Limits present-moment awareness and higher states of consciousness due to constant focus on potential dangers.	Reinforces neural pathways related to anxiety and fear, making it harder to trust and relax.
Reliance on External Authority	Hinders self-trust and personal autonomy, blocking access to inner wisdom and intuition.	Weakens self-reliance, creating a dependency on external reassurance that is hard to unlearn.
Emotional Reactivity & Anxiety	Keeps Type 6 in a heightened emotional state, reducing clarity and emotional regulation.	Strengthens anxious thought patterns, making it difficult to create new pathways for calm and resilience.
Conditioned Fear Responses	Narrows consciousness to defensive and reactive behaviors, limiting openness to new experiences.	Deeply ingrained fear responses make rewiring the brain for trust and openness a challenging process.
Need for External Validation	Prevents Type 6 from accessing higher self-confidence and trust in their own abilities.	Creates repetitive patterns of seeking approval, delaying the development of independent thinking.
Cycle of Anxiety & Worry	Keeps Type 6 trapped in low-frequency emotional states, preventing conscious growth toward peace.	Reinforces neural loops of overthinking and stress, requiring active practice to

Type 6: PRIMAL Rituals Recommended

Grounding rituals for Type 6s soothe their anxiety and fear by fostering a sense of safety within their body. These practices encourage vulnerability by helping them trust their instincts and let go of constant vigilance.

Open up to Your Breath

⇒ **Why**: Breathwork helps calm the nervous system, reducing anxiety and breaking cycles of fear.

⇒ **Frequency**: Daily, for 5–10 minutes.

Open up to Nature

⇒ **Why**: Spending time in nature grounds Type 6s, offering a sense of calm and security amidst uncertainty.

⇒ **Frequency**: Weekly.

Open up to Your Sleep

⇒ **Why**: Prioritizing sleep helps regulate emotions, restore mental clarity, and reduce stress-driven exhaustion.

⇒ **Frequency**: Nightly, for 7–8 hours.

Type 6: RATIONAL Rituals Recommended

RATIONAL rituals guide Type 6s to challenge their self-doubt and cultivate inner confidence. These practices reduce their reliance on external validation and build trust in their own decision-making abilities.

Open up to Laughter

⇒ **Why**: Humor reduces stress and shifts focus away from worry, fostering emotional resilience.

⇒ **Frequency**: Weekly through social interactions or lighthearted activities.

Open up to Learning

⇒ **Why**: Gaining knowledge in areas of uncertainty helps build confidence and reduce dependence on external validation.
⇒ **Frequency**: Regularly engage in skill-building or knowledge acquisition.

Open up to Gardening

⇒ **Why**: The repetitive and meditative nature of gardening provides stability and counteracts feelings of insecurity.
⇒ **Frequency**: Weekly.

Type 6: SPIRITUAL Rituals Recommended

SPIRITUAL rituals for Type 6s foster faith and connection to a higher power. By embracing surrender and letting go of fear, they deepen their sense of security and align with their spiritual purpose.

Open up to Meditation

⇒ **Why**: Mindfulness meditation calms anxiety-driven thoughts, fostering trust in oneself and inner peace.
⇒ **Frequency**: Daily, for 10–15 minutes.

Open up to Prayers

⇒ **Why**: Prayer or reflective connection with a higher power alleviates fears and reinforces inner security.
⇒ **Frequency**: Daily.

Type 6: Compatibility With Other Types

Type 1 (The Reformer)

•**What Works**: Both are responsible and committed to doing what's right. Type 1's structure provides stability, and Type 6's loyalty reinforces trust. Together, they create a reliable and organized team.
•**Potential Friction**: Type 1's perfectionism can increase Type 6's anxiety, leading to tension. Type 6's indecision may frustrate Type 1's desire for quick resolution.

Type 2 (The Helper)

•**What Works**: Type 2's nurturing reassures Type 6's need for security, and Type 6's loyalty provides a stable foundation for Type 2's giving nature. They both enjoy supporting one another emotionally.
•**Potential Friction**: Type 6's anxiety may make Type 2 feel underappreciated if they constantly seek reassurance. Type 2's need for affection could overwhelm a cautious Type 6.

Type 3 (The Achiever)

•**What Works**: Type 6's support and reliability allow Type 3 to feel secure while pursuing their goals. Type 6 appreciates Type 3's ability to take decisive action.
•**Potential Friction**: Type 3's focus on success may leave Type 6 feeling uncertain or neglected, while Type 3 could view Type 6's hesitancy as a hindrance to their ambitions.

Type 4 (The Individualist)

•**What Works**: Both types desire meaningful connections, with Type 6 providing stability and Type 4 offering emotional depth. Together, they create a strong emotional bond.
•**Potential Friction**: Type 4's emotional volatility and introspection can increase Type 6's fears and need for reassurance. Type 6's desire for security may feel restrictive to a free-spirited Type 4.

Type 5 (The Investigator)

•**What Works**: Both types value knowledge and thinking things through. Type 5's intellectual approach helps Type 6 feel more secure in decisions, and Type 6's loyalty offers comfort to Type 5's detachment.
•**Potential Friction**: Type 6's need for reassurance can feel intrusive to independent Type 5. Type 5's tendency to withdraw may increase Type 6's anxiety or sense of insecurity.

Type 6 (The Loyalist)

•**What Works**: Two Type 6s together can create a relationship based on deep loyalty and mutual understanding. They both appreciate commitment and working through challenges together.
•**Potential Friction**: Their shared tendency toward anxiety may amplify fears and indecision. Both might look to each other for reassurance without resolving their own doubts, leading to a cycle of worry.

Type 7 (The Enthusiast)

•**What Works**: Type 7 brings energy and optimism, helping to lift Type 6 out of anxiety. Type 6's grounded nature can help Type 7 focus and feel secure.
•**Potential Friction**: Type 7's need for excitement and spontaneity may clash with Type 6's desire for security and predictability. Type 6 may feel unsupported if Type 7 avoids dealing with serious issues.

Type 8 (The Challenger)

•**What Works**: Type 8's confidence provides protection and security for Type 6, while Type 6's loyalty fosters trust in Type 8. Together, they form a strong, protective bond.
•**Potential Friction**: Type 8's direct and aggressive nature can overwhelm Type 6, leading to heightened anxiety. Type 6's indecision may frustrate Type 8's need for quick, decisive action.

Type 9 (The Peacemaker)

•**What Works**: Both types value stability and security. Type 6 appreciates Type 9's calmness and non-confrontational nature, while Type 9 enjoys Type 6's commitment and loyalty.
•**Potential Friction**: Type 6's tendency to overthink can disrupt the peace Type 9 seeks. Type 9's avoidance of conflict may increase Type 6's anxiety, leading to a lack of resolution.

Type 6: Compatibility Table

Pair	Compatibility Dynamics	Strengths Together	Challenges Together	Tips for Improving Flow
6 and 1	Both value responsibility and structure, creating a solid foundation for trust and mutual respect.	Shared dedication to doing the right thing and creating stability.	Type 6's anxiety can clash with Type 1's perfectionism, leading to over-analysis and stress.	Focus on celebrating progress rather than seeking perfection to ease mutual tensions.
6 and 2	A supportive and nurturing dynamic where Type 2 provides warmth and reassurance.	Strong emotional connection and mutual care for loved ones.	Type 6's need for reassurance may overwhelm Type 2, who might prioritize others' needs excessively.	Maintain open communication about boundaries and ensure mutual emotional support.
6 and 3	Type 3's confidence complements Type 6's cautious approach, creating a balanced team dynamic.	Shared goals and complementary strengths in planning and execution.	Type 6's doubts can conflict with Type 3's fast-paced drive, creating friction around decision-making.	Encourage Type 3 to slow down and involve Type 6 in planning to build confidence and alignment.

Pair	Compatibility Dynamics	Strengths Together	Challenges Together	Tips for Improving Flow
6 and 4	Emotional depth from Type 4 complements Type 6's loyalty, fostering meaningful connections.	Strong emotional bonds and a shared appreciation for authenticity.	Type 6's need for stability may feel stifling to Type 4's desire for individuality and emotional exploration.	Find a balance between stability and creative freedom, ensuring mutual trust and respect for differences.
6 and 5	Type 5's analytical approach balances Type 6's emotional intensity, creating a thoughtful partnership.	Shared love for learning and problem-solving with practical outcomes.	Type 6's need for connection may conflict with Type 5's preference for solitude and independence.	Schedule regular check-ins to ensure both partners feel secure without overwhelming each other.
6 and 6	Shared values of loyalty and security foster a deep, trusting relationship.	Strong mutual support and a shared desire for a stable, predictable environment.	Amplified anxiety and overthinking can create loops of fear-based decision-making.	Practice mindfulness and intentionally balance planning with moments of relaxation and spontaneity.
6 and 7	Type 7's optimism counters Type 6's worry, creating a dynamic mix of caution and adventurous energy.	Complementary energy where Type 7 brings excitement and Type 6 ensures safety.	Type 6's cautious approach may feel restrictive to Type 7's free-spirited nature, leading to frustration.	Emphasize collaborative decision-making and balance risk-taking with careful planning.

Pair	Compatibility Dynamics	Strengths Together	Challenges Together	Tips for Improving Flow
6 and 8	Type 8's strength provides reassurance to Type 6, while Type 6's loyalty fosters trust with Type 8.	A dynamic duo where Type 8 provides leadership and Type 6 ensures stability and foresight.	Type 6's self-doubt may clash with Type 8's assertiveness, creating feelings of intimidation or tension.	Build mutual understanding through open dialogue and allow Type 6 to voice concerns without judgment.
6 and 9	Type 9's calm demeanor soothes Type 6's anxiety, fostering harmony and balance.	A peaceful dynamic where Type 6 feels reassured and Type 9 enjoys a steady partnership.	Type 6's overthinking may disrupt Type 9's desire for peace, leading to avoidance or suppressed emotions.	Encourage Type 9 to engage in discussions while helping Type 6 ease worries through mindful reassurance.

Type 7: The Enthusiast (Energetic, Adventurous, Optimistic, Spontaneous)

Energetic and optimistic, Enthusiasts are driven by a desire to seek joy, variety, and freedom. They thrive on exploration and new experiences, often bringing a sense of excitement and possibility to their environments.

Each type develops unique coping mechanisms—patterns of thought, emotion, and behavior formed to address their core fears and navigate life's challenges. For Enthusiasts, this often involves avoiding pain or discomfort by staying busy and focusing on positive possibilities. While this fosters adaptability and creativity, it can lead to overcommitment or escapism.

In the following sections, we explore how Type 7 typically operates in key areas of life—at work, home, and in relationships—revealing their characteristic strengths, vulnerabilities, and tendencies. We also examine this type's consciousness and plasticity challenges, identifying opportunities to enhance self-awareness, adaptability, and emotional intelligence. To support these efforts, we provide tailored *OPEN rituals*—PRIMAL, RATIONAL, and SPIRITUAL practices that nurture openness, balance, and personal growth.

Finally, the section concludes by unpacking compatibility dynamics, offering insights into how Type 7 relates to all other Enneagram types across various professional, familial, or romantic relationships.

The Gifts of Being an Enneagram 7

The gifts of an Enneagram Type 7, "The Enthusiast," are rooted in their zest for life, boundless curiosity, and ability to bring joy and inspiration to others. Here's a deeper look at these gifts:

1. Optimism and Positivity: Type 7s are natural optimists. They can see the bright side of any situation, bringing light and positivity even under challenging circumstances. Their enthusiasm is contagious, and they can lift the spirits of those around them, making them the go-to people for encouragement and motivation.

2. Adventure and Curiosity: Sevens are explorers driven by an insatiable curiosity about the world. Whether seeking new ideas, experiences, or relationships, they bring a spirit of adventure that opens doors to unexpected opportunities.

3. Creativity and Innovation: With their open-minded approach and hunger for new experiences, Type 7s are highly creative individuals. They are often brimming with ideas and can think outside the box, offering fresh perspectives and solutions in both work and personal life.

4. Playfulness and Fun: Type 7s bring an infectious sense of fun to everything they do. They enjoy life, finding humor and playfulness even in routine activities. This makes them exciting companions who turn ordinary moments into adventures.

5. Resilience and Flexibility: Sevens quickly bounce back from challenges. Their natural inclination to focus on the future and what's possible rather than dwelling on setbacks allows them to remain resilient in difficult times. They adapt quickly and look for new paths forward when obstacles arise.

6. Visionary Thinking: Type 7s are often seen as visionaries. They can imagine future possibilities, inspiring others with their forward-thinking and big-picture mindset. They dream of what could be and often see the potential that others miss.

7. Inspiring and Empowering Others: Sevens have an infectious energy that motivates others to explore their potential. They are encouraging and help people see the world as full of possibilities. Their enthusiastic approach empowers others to break free from their limitations and explore new opportunities.

8. Capacity for Joy and Gratitude: Sevens have an immense capacity for joy, often reminding those around them of the importance of savoring life's pleasures. Whether through adventure, relationships, or personal growth, they delight in a wide range of experiences.

9. Multitasking and Versatility: Type 7s are excellent multitaskers, often juggling several projects or ideas simultaneously. Their love for variety and ability to adapt to different circumstances means they can excel in diverse environments and bring a wealth of experience to any situation.

The gifts of a Type 7 lie in their ability to bring light, energy, and enthusiasm into every space they enter. Their adventurous spirit, creativity, and ability to inspire others make them natural leaders and visionaries. They bring a sense of possibility, joy, and innovation, helping others expand their horizons and see life through a more optimistic, vibrant lens. However, the true beauty of Type 7's gifts shines brightest when they balance their innate positivity with the ability to stay grounded and present in both joy and discomfort, leading to a more prosperous, more integrated life.

Type 7 Coping Mechanisms

Type 7s, the Enthusiasts, often cope by seeking excitement and avoiding discomfort. They are naturally drawn to adventure, novelty, and fun, using these experiences to escape emotional pain or limitations. While this helps them maintain a sense of joy and optimism, it can also lead to distraction, overindulgence, and difficulty facing deeper emotions.

1. Avoidance of Pain and Discomfort

- **Coping Mechanism**: As children, Type 7s learned to divert their attention from painful or boring situations by focusing on pleasurable experiences. This habit helped them feel in control of their emotional state.
- **Adult Challenge**: In adulthood, this avoidance can lead to difficulty processing difficult emotions. They may struggle with sitting in discomfort, often jumping from one activity to another instead of addressing underlying pain or challenges.

2. Escaping Through Stimulation and Distraction

- **Coping Mechanism**: Type 7s cope by keeping their minds occupied with new experiences, hobbies, and projects. Their constant search for stimulation helped them avoid feelings of emptiness or dissatisfaction.
- **Adult Challenge**: This can make it hard to commit to long-term goals or relationships. They may struggle with follow-through, often abandoning tasks or responsibilities when they become routine or unexciting.

3. Reframing Negative Experiences

- **Coping Mechanism**: To avoid being overwhelmed by difficulties, Type 7s developed the ability to reframe challenges into something positive. They find ways to see the silver lining in negative experiences.
- **Adult Challenge**: While this optimism is a strength, it can also lead to denial of real problems. Instead of facing struggles head-on, Type 7s may dismiss or downplay important issues, missing opportunities for growth.

4. Overindulgence and Escapism

- **Coping Mechanism**: Seeking pleasure became a way for Type 7s to fill emotional voids. They may turn to food, shopping, travel, or social events to avoid dealing with stress.
- **Adult Challenge**: This tendency can lead to impulsive behaviors, financial instability, or unhealthy habits. Overindulgence can become a substitute for deeper emotional fulfillment, making it difficult for them to experience true satisfaction.

5. Focusing on the Future to Avoid the Present

- **Coping Mechanism**: Type 7s often focus on what's next, planning future adventures or ideas to avoid feeling trapped in the present. This habit keeps them excited and hopeful.
- **Adult Challenge**: This can prevent them from fully engaging in the present moment. They may struggle with mindfulness, always chasing the next thrill instead of appreciating what they have now.

6. Minimizing Emotional Depth

- **Coping Mechanism**: Type 7s use humor, charm, and quick thinking to keep interactions light. They avoid emotional depth because they fear getting stuck in negative emotions.
- **Adult Challenge**: This can make it difficult to build deep, meaningful relationships. Others may see them as fun but emotionally unavailable, leading to feelings of loneliness or superficial connections.

7. Rationalizing Avoidance

- **Coping Mechanism**: To justify avoiding painful experiences, Type 7s rationalize their choices, convincing themselves that moving on is the best option.
- **Adult Challenge**: This mental reframing can prevent them from addressing unresolved emotions or conflicts, leading to patterns of avoidance in relationships, work, or personal growth.

By recognizing these coping mechanisms, Type 7s can work toward finding a balance between excitement and emotional depth. Learning to stay present, embrace discomfort, and cultivate deeper connections will help them experience a more fulfilling and grounded life.

Type 7 at Work

Type 7s bring energy, creativity, and a sense of adventure to their professional life. They thrive in environments that value innovation and flexibility but may struggle with routines or tasks requiring sustained focus.

Strengths

⇒ **Energetic and Motivating**: Type 7s bring excitement and optimism, uplifting the morale of their teams and fostering a positive work environment.

- ⇒ **Creative Thinkers**: They excel at brainstorming and generating innovative solutions, thriving in roles that value outside-the-box thinking.
- ⇒ **Adaptable and Resilient**: Type 7s easily handle change and uncertainty and often thrive in fast-paced or dynamic workplaces.
- ⇒ **Visionary Approach**: They are forward-thinking and enjoy exploring new possibilities, making them great at identifying growth opportunities

Challenges

- ⇒ **Difficulty with Focus**: Their craving for variety may lead to scattered efforts or incomplete projects.
- ⇒ **Avoidance of Routine**: They often resist repetitive or detail-heavy tasks, hindering long-term productivity.
- ⇒ **Overcommitment**: Type 7s may take on too many responsibilities, leading to burnout or failure to meet expectations.
- ⇒ **Avoidance of Conflict**: Their focus on positivity may cause them to shy away from addressing complicated issues or providing critical feedbackType 6 at Home

Type 7 at Home

Type 7s bring a lively and adventurous spirit to their home environment. They are often the driving force behind fun activities and spontaneous plans, making their home vibrant. However, their need for stimulation can sometimes lead to responsibilities and emotional depth challenges.

Strengths

- ⇒ **Joyful and Fun**: Type 7s infuse their home life with energy and excitement, often creating memorable experiences for their family or housemates.
- ⇒ **Adventurous Spirit**: They enjoy planning trips or trying new activities, keeping life at home dynamic and engaging.

⇒ **Positive Energy**: Their optimism helps create a lighthearted and uplifting environment, even in stressful times.

Challenges

⇒ **Avoidance of Responsibility**: Type 7s may struggle with routine household tasks or emotionally heavy conversations, preferring to focus on enjoyable activities.
⇒ **Disengagement**: Their desire for constant stimulation can lead to neglecting necessary but mundane responsibilities.
⇒ **Shallow Focus**: They may avoid addressing deeper emotional issues, which could create distance in close relationships.

Type 7 in Relationships

Type 7s bring excitement and enthusiasm to relationships, often creating vibrant and adventurous dynamics. However, their need for freedom and avoidance of emotional discomfort can challenge sustaining deeper connections. Below is a breakdown of their strengths and challenges in relationships:

Strengths

⇒ **Charming and Fun**: Type 7s are engaging and spontaneous, making relationships with them full of adventure and joy.
⇒ **Optimism**: They help their partners focus on the brighter side of life, lifting spirits during tough times.
⇒ **Curiosity and Exploration**: They are eager to try new things and explore shared interests, adding variety and excitement to the relationship.

Challenges

⇒ **Fear of Commitment**: Their desire for freedom and aversion to limitations can make committing to long-term emotional depth challenging.

⇒ **Avoidance of Emotional Pain**: Type 7s may avoid addressing conflicts or deeper issues, focusing instead on lighter topics.

⇒ **Restlessness**: Their constant pursuit of new experiences can leave partners feeling neglected or overshadowed by their need for stimulation.

Type 7 Consciousness Challenges

Type 7s, the enthusiastic and adventurous "Enthusiasts," often avoid discomfort by seeking constant stimulation and new experiences. This tendency to focus on external pleasures can make it difficult for them to sit with challenging emotions or fully engage with the present moment. Their avoidance of more profound emotional experiences may hinder their ability to develop self-awareness and cultivate a more grounded and reflective consciousness.

1. Avoiding Present-Moment Awareness

Sevens naturally avoid staying in the present, especially if it involves discomfort or emotional pain. They focus on future possibilities and exciting plans to escape uncomfortable feelings. This avoidance prevents them from engaging in the self-awareness and mindfulness necessary for deeper consciousness. For true personal growth, Sevens must learn to embrace the present moment, even when challenging or less stimulating.

2. Escaping Through External Stimulation

To avoid emotional discomfort, Sevens fill their lives with activities and distractions. This constant pursuit of external stimulation means they often overlook deeper emotional truths, limiting their ability to reflect and raise consciousness. They avoid internal stillness by continuously moving from one distraction to another, critical for consciousness expansion.

4. Fear of Emotional Depth

Type 7s avoid emotional vulnerability because they fear being overwhelmed by negative feelings. As a result, they shy away from exploring complicated feelings. Developing greater consciousness requires them to engage with their emotions, not simply reframe everything in a positive light. Without addressing emotional depth, they remain stuck at the surface level of awareness.

5. Superficial Engagement in Mindfulness Practices

The fast-paced nature of Type 7's mind makes it difficult to settle into practices like meditation or introspection, essential for raising consciousness. Their desire for excitement and novelty can make mindfulness practices dull or constraining. However, these practices are critical for developing self-awareness and emotional resilience, providing the stillness needed for conscious reflection.

6. Difficulty Embracing Stillness

Sevens often struggle to sit in stillness or engage in reflective activities that don't immediately excite them. Consciousness requires periods of quiet reflection and presence with oneself, but this can feel stifling to a Type 7, who thrives on movement and action. Learning to value stillness as a source of growth is essential for raising their consciousness.

Type 7: Plasticity Challenges

Neuroplasticity refers to the brain's ability to reorganize itself by forming new neural connections throughout life. Sevens can change habits, beliefs, and emotional responses through new experiences and learning.

1. Scattered Focus

Sevens are naturally drawn to novelty and excitement, but their scattered attention makes engaging in sustained, focused learning difficult. Plasticity requires repeated engagement with a habit or skill to create new neural pathways, but Sevens often move on to the next exciting thing before real change can occur. This prevents them from building the deep neural connections needed for personal transformation.

2. Avoidance of Emotional Discomfort

Sevens often avoid habits that require them to face uncomfortable emotions, such as mindfulness or therapy. Effective plasticity involves rewiring emotional and cognitive patterns, but this process can involve mental discomfort, which Sevens are quick to avoid. As a result, they may miss opportunities to create new, healthier patterns because they shy away from the necessary emotional work.

3. Impulsivity and Short-Term Focus

While Sevens are highly adaptable, their impulsivity often disrupts long-term focus. Plasticity thrives on consistent practice, but Sevens may abandon new habits once the novelty wears off. This short-term focus prevents the brain from reinforcing new pathways and turning those habits into lasting change. Cultivating patience and consistency is critical for Sevens who want to benefit from their brain's adaptability.

4. Overcommitment to New Ideas

Sevens often overcommit to multiple projects, ideas, or social engagements, which can overwhelm their capacity for focused attention. This scattered energy makes it difficult for them to stick with a single habit or practice long enough to see meaningful change. Brain plasticity requires sustained focus on particular areas of growth, and the challenge for Sevens is learning to commit to fewer, more meaningful changes.

5. Disconnection Between Emotional and Cognitive Growth

Sevens are quick thinkers and eager learners, but they often focus on cognitive growth at the expense of emotional development. True plasticity involves both emotional and cognitive rewiring, but Sevens may avoid the emotional aspects of growth, focusing instead on the intellectual stimulation of learning new things. Integrating both cognitive and emotional development is key to achieving meaningful plasticity.

Type 7: Summary of Challenges

Challenge	Impact on Consciousness	Impact on Plasticity
Avoiding Present-Moment Awareness	Limits self-awareness and mindfulness; prevents deep reflection on emotions.	Hinders the development of consistent habits that foster long-term change.
Escaping Through Stimulation	Keeps focus external, avoiding emotional growth and inner stillness.	Leads to scattered focus, making it difficult to reinforce new neural pathways.
Fear of Emotional Depth	Prevents engagement with difficult emotions, limiting self-insight and emotional resilience.	Avoids discomfort required for emotional rewiring, missing opportunities for deep change.
Superficial Engagement in Mindfulness	Struggles to practice stillness, limiting the expansion of consciousness and self-awareness.	Impulsivity leads to lack of repetition, preventing sustained neural rewiring.
Difficulty Embracing Stillness	Resists introspection, hindering spiritual growth and present-moment consciousness.	Avoids long-term practice needed to strengthen neural pathways for emotional resilience.

Type 7: PRIMAL Rituals Recommended

For Type 7s, primal rituals ground their restless energy and connect them to the present moment. These practices soften their avoidance of discomfort and encourage vulnerability by fostering presence and mindfulness.

Open up to Your Breath

⇒ **Why**: Helps Type 7s stay grounded in the present moment, regulating their tendency to avoid emotional discomfort.
⇒ **Frequency**: Daily for 5–10 minutes.

Open up to Your Sleep

⇒ **Why**: Prioritizing quality sleep helps consolidate memories and reduce stress, promoting overall emotional resilience.
⇒ **Frequency**: Every night with a consistent bedtime routine.

Open up to Nature

⇒ **Why**: Spending time outdoors reduces stress and improves cognitive clarity, providing relaxation from overstimulation.
⇒ **Frequency**: Weekly, for 30–60 minutes.

Open up to Your Food

⇒ **Why**: Mindful eating reconnects Type 7s with their body, promoting balance and present-moment awareness to stabilize mood.
⇒ **Frequency**: Daily during meals.

Type 7: RATIONAL Rituals Recommended

RATIONAL rituals for Type 7s encourage self-reflection and emotional balance. Embracing stillness and deeper thinking, these practices help them confront their fears and reduce their need to escape through distraction.

Open up to Laughter

⇒ **Why**: Laughter boosts endorphins, reduces cortisol, and deepens joy-related neural pathways.

⇒ **Frequency**: Weekly through social interactions or lighthearted activities.

Open up to Music

⇒ **Why**: Engaging with evocative music stimulates emotional and cognitive centers, providing a focus for emotional release.

⇒ **Frequency**: Weekly or during moments of reflection.

Open up to Cooking

⇒ **Why**: Cooking encourages mindfulness through sensory engagement, helping Type 7s stay grounded and focused.

⇒ **Frequency**: Weekly as a creative, meditative activity.

Open up to Learning

⇒ **Why**: Exploring new topics fosters cognitive flexibility and long-term neural adaptability.

⇒ **Frequency**: Regularly, 1–2 hours per week.

Type 7: SPIRITUAL Rituals Recommended

SPIRITUAL rituals for Type 7s invite them to explore contentment and surrender. Through meditation or prayer, they can nurture gratitude and discover fulfillment beyond external pleasures.

Open up to Meditation

⇒ **Why**: Calms a restless mind, reduces distracting thoughts, and fosters emotional regulation.

⇒ **Frequency**: Daily for 10–15 minutes.

Open up to Rituals

⇒ **Why**: Rituals provide mindfulness and structure, helping Type 7s embrace stillness and present-moment awareness.
⇒ **Frequency**: Weekly as a grounding practice.

Open up to Stargazing

⇒ **Why**: Promotes awe and tranquility, helping Type 7s transcend daily distractions and connect spiritually.
⇒ **Frequency**: Monthly or as needed.

Open up to Mantras

⇒ **Why**: Repeating mantras helps focus attention, calm the mind, and enhance mindfulness.
⇒ **Frequency**: Daily during meditation or quiet reflection.

Type 7: Compatibility With Other Types

Type 1 (The Reformer)

•**What Works**: Both are responsible and committed to doing what's right. Type 1's structure provides stability, and Type 6's loyalty reinforces trust. Together, they create a reliable and organized team.

•**Potential Friction**: Type 1's perfectionism can increase Type 6's anxiety, leading to tension. Type 6's indecision may frustrate Type 1's desire for quick resolution.

Type 2 (The Helper)

•**What Works**: Type 2's nurturing reassures Type 6's need for security, and Type 6's loyalty provides a stable foundation for Type 2's giving nature. They both enjoy supporting one another emotionally.

•**Potential Friction**: Type 6's anxiety may make Type 2 feel underappreciated if they constantly seek reassurance. Type 2's need for affection could overwhelm a cautious Type 6.

Type 3 (The Achiever)

•**What Works**: Type 6's support and reliability allow Type 3 to feel secure while pursuing their goals. Type 6 appreciates Type 3's ability to take decisive action.
•**Potential Friction**: Type 3's focus on success may leave Type 6 feeling uncertain or neglected, while Type 3 could view Type 6's hesitancy as a hindrance to their ambitions.

Type 4 (The Individualist)

•**What Works**: Both types desire meaningful connections, with Type 6 providing stability and Type 4 offering emotional depth. Together, they create a strong emotional bond.
•**Potential Friction**: Type 4's emotional volatility and introspection can increase Type 6's fears and need for reassurance. Type 6's desire for security may feel restrictive to a free-spirited Type 4.

Type 5 (The Investigator)

•**What Works**: Both types value knowledge and thinking things through. Type 5's intellectual approach helps Type 6 feel more secure in decisions, and Type 6's loyalty offers comfort to Type 5's detachment.
•**Potential Friction**: Type 6's need for reassurance can feel intrusive to independent Type 5. Type 5's tendency to withdraw may increase Type 6's anxiety or sense of insecurity.

Type 6 (The Loyalist)

•**What Works**: Two Type 6s can create a relationship based on deep loyalty and mutual understanding. They both appreciate the commitment and working through challenges together.

•**Potential Friction**: Their shared tendency toward anxiety may amplify fears and indecision. Both might look to each other for reassurance without resolving their own doubts, leading to a cycle of worry.

Type 7 (The Enthusiast)

•**What Works**: Type 7 brings energy and optimism, helping to lift Type 6 out of anxiety. Type 6's grounded nature can help Type 7 focus and feel secure.
•**Potential Friction**: Type 7's need for excitement and spontaneity may clash with Type 6's desire for security and predictability. Type 6 may feel unsupported if Type 7 avoids dealing with serious issues.

Type 8 (The Challenger)

•**What Works**: Type 8's confidence provides protection and security for Type 6, while Type 6's loyalty fosters trust in Type 8. Together, they form a strong, protective bond.
•**Potential Friction**: Type 8's direct and aggressive nature can overwhelm Type 6, leading to heightened anxiety. Type 6's indecision may frustrate Type 8's need for quick, decisive action.

Type 9 (The Peacemaker)

•**What Works**: Both types value stability and security. Type 6 appreciates Type 9's calmness and non-confrontational nature, while Type 9 enjoys Type 6's commitment and loyalty.
•**Potential Friction**: Type 6's tendency to overthink can disrupt the peace Type 9 seeks. Type 9's avoidance of conflict may increase Type 6's anxiety, leading to a lack of resolution.

Type 7: Compatibility Table

Pair	Compatibility Dynamics	Strengths Together	Challenges Together	Tips for Improving Flow
7 and 1	A structured, disciplined partner balances Type 7's spontaneity, creating a dynamic of planning vs. exploration.	Type 1 grounds Type 7's adventurous spirit; Type 7 brings joy and optimism to Type 1's focus on improvement	Type 1 may view Type 7 as scattered or irresponsible; Type 7 may feel restricted by Type 1's rigidity.	Foster mutual respect: Type 7 should honor Type 1's need for structure, while Type 1 embraces spontaneity.
7 and 2	A nurturing, empathetic partner complements Type 7's adventurous nature, fostering mutual excitement for shared experiences.	Type 2 offers emotional depth; Type 7 keeps the relationship lively and fun.	Type 2 may feel neglected if Type 7 focuses too much on external pursuits; Type 7 may feel smothered by Type 2's care.	Create balance: Type 7 should prioritize time with Type 2, while Type 2 allows Type 7 freedom to explore.
7 and 3	Both partners are ambitious and enthusiastic, creating a dynamic of shared goals and mutual motivation.	Both value energy and excitement, excelling together in achieving big dreams.	Competitive tendencies may arise; Type 3's work focus may clash with Type 7's desire for play.	Set joint goals: Prioritize activities that merge work and leisure to maintain harmony and flow.

Pair	Compatibility Dynamics	Strengths Together	Challenges Together	Tips for Improving Flow
7 and 4	Type 4's introspection contrasts with Type 7's outward energy, creating a balance between depth and lightheartedness.	Type 4 encourages emotional reflection; Type 7 provides lightness and joy.	Type 7 may find Type 4 too intense or moody; Type 4 may see Type 7 as superficial or avoidant.	Embrace differences: Type 7 should practice patience with Type 4's emotional needs; Type 4 can enjoy Type 7's playfulness.
7 and 5	Type 5's intellectual depth complements Type 7's adventurous spirit, fostering a dynamic of curiosity and discovery.	Type 5 encourages thoughtful exploration; Type 7 energizes and inspires Type 5.	Type 7 may see Type 5 as withdrawn or too reserved; Type 5 may feel drained by Type 7's high energy.	Balance energy: Type 7 should slow down for Type 5's pace; Type 5 can occasionally join Type 7's adventures.
7 and 6	Type 6 offers grounding and loyalty, which stabilizes Type 7's adventurous tendencies, creating a balance of security and fun.	Type 6 provides support and stability; Type 7 infuses excitement and optimism into the relationship.	Type 6 may worry about Type 7's impulsiveness; Type 7 may feel held back by Type 6's cautious nature.	Build trust: Type 7 should reassure Type 6; Type 6 can embrace occasional risk-taking with Type 7.

Pair	Compatibility Dynamics	Strengths Together	Challenges Together	Tips for Improving Flow
7 and 7	Both partners share a love for excitement and variety, creating a highly energetic and dynamic relationship.	Mutual optimism and enthusiasm for new experiences strengthen their bond.	A lack of grounding may lead to neglecting responsibilities or avoiding challenges.	Create structure: Incorporate rituals or shared responsibilities to balance fun with focus.
7 and 8	Type 8's assertiveness energizes Type 7's adventurous side, creating a dynamic of boldness and exploration.	Type 8 brings strength and decisiveness; Type 7 contributes playfulness and optimism.	Type 8 may feel Type 7 is unreliable; Type 7 may view Type 8 as overbearing.	Find balance: Type 7 can commit to plans; Type 8 can practice flexibility and humor.
7 and 9	Type 9's calm and harmonious nature complements Type 7's lively energy, fostering a relationship of peace and excitement.	Type 9 stabilizes Type 7's enthusiasm; Type 7 inspires Type 9 to embrace new experiences.	Type 9 may become passive or avoidant; Type 7 may overlook Type 9's deeper needs.	Communicate openly: Type 7 should listen to Type 9's desires; Type 9 can actively participate in Type 7's plans.

WHICH RITUALS ARE RIGHT FOR YOU?

Rituals are structured activities designed to engage the mind, body, and spirit in ways that promote healing, growth, and flow. The *OPEN framework* identifies PRIMAL, RATIONAL, and SPIRITUAL rituals as distinct yet complementary practices that engage different aspects of human experience and consciousness.

PRIMAL rituals are deeply rooted in physical and sensory engagement with the world, often designed to connect individuals with their instinctual nature. These rituals might involve physical activity, connection with natural elements, or experiences that invoke a sense of grounding and vitality. Examples include walking in nature or engaging in rhythmic dance. PRIMAL rituals are essential for resetting your nervous systems, fostering resilience, and connecting with your body.

RATIONAL rituals, on the other hand, are structured practices that calm the mind and evoke emotional satisfaction. Practices such as learning an instrument and engaging in artistic expression allow individuals to explore creativity while providing a sense of accomplishment and balance. RATIONAL rituals bridge the gap between instinct and higher consciousness by cultivating emotional resilience and cognitive clarity.

Finally, SPIRITUAL rituals delve into transcendence, purpose, and connection with a higher state of being. These practices, such as meditation, prayer, or chanting mantras, help individuals achieve inner peace and cultivate a sense of unity with the greater universe. SPIRITUAL rituals nurture the soul, promoting introspection and a more profound sense of fulfillment. Together, these rituals support a holistic approach to well-being, allowing individuals to balance instinct, intellect, and spirit.

These foundational practices ensure that the transformative power of rituals unfolds in your life, helping you rewire maladaptive patterns and align with your highest potential. Let this guide support you as you create a sustainable routine filled with meaning and flow. Pair your practice with mantras to reinforce positive intentions, and approach yourself with self-compassion, recognizing that every step forward is progress. Regularity and commitment are essential—what matters most is showing up consistently. Additionally, set your intention (your "why") before each ritual and choose a setting (your "where") that feels safe, calming, and inspiring.

The following tables offer a curated starting point for each type to choose rituals tailored to their unique strengths, challenges, and core motivations. A ritual doesn't have to be elaborate or time-consuming; even a few minutes daily can create a meaningful shift in your neurospiritual journey.

PRIMAL Rituals Most Important for each Type

Ritual	Description	Types That Benefit Most	Why
Open up to Your Body (see Yoga, dancing and running below)	Practice body awareness through movement and somatic exercises to release stress and tension.	Type 8 (The Challenger), Type 9 (The Peacemaker)	Grounds Type 8's energy; supports Type 9's need for balance and calm.
Open up to Yoga	Engage in yoga practice for flexibility, mindfulness, and strength.	Type 5 (The Investigator), Type 4 (The Individualist)	Promotes emotional balance for Type 4; fosters focus and calm introspection for Type 5.
Open up to Dancing	Express joy and release tension through dance as a creative movement form.	Type 7 (The Enthusiast), Type 2 (The Helper)	Channels Type 7's exuberance provides emotional expression for Type 2.

Ritual	Description	Types That Benefit Most	Why
Open up to Your Breath	Practice breathwork to calm the nervous system and enhance emotional regulation.	Type 1 (The Reformer), Type 9 (The Peacemaker)	Reduces stress in perfection-driven Type 1; eases inertia for Type 9.
Open up to Your Sleep	Cultivate restorative sleep habits for emotional and cognitive resilience.	Type 6 (The Loyalist), Type 3 (The Achiever)	Helps Type 6 manage anxiety-induced insomnia; aids Type 3 in sustaining energy for goals.
Open up to Nature	Spend time outdoors to promote grounding, stress reduction, and connection with the environment.	Type 4 (The Individualist), Type 9 (The Peacemaker)	Grounds Type 4's emotional highs; calms and harmonizes Type 9.
Open up to Your Food	Embrace mindful eating and nutritional balance to support well-being.	Type 2 (The Helper), Type 3 (The Achiever)	Encourages self-nourishment for Type 2; sustains energy for ambitious Type 3.
Open up to Your Pets	Develop bonds with pets to foster emotional connection and relieve stress.	Type 2 (The Helper), Type 6 (The Loyalist)	Offers unconditional love for Type 2; builds trust and calms anxieties for Type 6.

RATIONAL Rituals Most Important for each Type

Ritual	Description	Types That Benefit Most	Why
Open up to Evocative Music	Listen to meaningful music to evoke deep emotions and improve mood.	Type 4 (The Individualist), Type 9 (The Peacemaker)	Resonates with Type 4's depth; relaxes and harmonizes Type 9.
Open up to an Instrument	Learn or play a musical instrument to boost creativity and focus.	Type 5 (The Investigator), Type 7 (The Enthusiast)	Encourages exploration for Type 5; satisfies Type 7's need for novelty.
Open up to Laughter	Practice humor to reduce stress and elevate mood.	Type 7 (The Enthusiast), Type 2 (The Helper)	Energizes Type 7; lightens Type 2's intense emotional focus.
Open up to Uplifting Movies	Watch inspiring films to generate hope and positive emotions.	Type 3 (The Achiever), Type 6 (The Loyalist)	Inspires Type 3's ambition; reassures Type 6 by reinforcing hope.
Open up to Gardening	Engage in planting and nurturing to foster mindfulness and grounding.	Type 9 (The Peacemaker), Type 4 (The Individualist)	Grounds Type 9 in peaceful routine; nurtures creativity for Type 4.
Open up to Cooking	Create meals as a form of creative expression and care.	Type 6 (The Loyalist), Type 2 (The Helper)	Provides stability for Type 6; channels Type 2's desire to nurture others.

Ritual	Description	Types That Benefit Most	Why
Open up to Board Games	Play games that encourage strategic thinking and social connection.	Type 3 (The Achiever), Type 5 (The Investigator)	Appeals to Type 3's competitive nature; stimulates Type 5's intellect.
Open up to Non-violent Video Games	Engage in gameplay to develop strategy and relaxation without conflict.	Type 7 (The Enthusiast), Type 9 (The Peacemaker)	Adds joy for Type 7; provides relaxation for Type 9.
Open up to Learning	Pursue education or skill acquisition to engage curiosity.	Type 5 (The Investigator), Type 3 (The Achiever)	Satisfies intellectual needs for Type 5; supports goals for Type 3.
Open up to Traveling	Explore new places to expand perspective and inspiration.	Type 7 (The Enthusiast), Type 4 (The Individualist)	Fulfills adventure-seeking in Type 7; inspires creativity in Type 4.
Open up to Art	Create or engage with visual art to express emotions and foster creativity.	Type 4 (The Individualist), Type 5 (The Investigator)	Offers an outlet for Type 4's introspection; provides exploration for Type 5.

SPIRITUAL Rituals Most Important for each Type

Ritual	Description	Types That Benefit Most	Why
Open up to Meditation	Engage in mindfulness meditation for emotional regulation and inner peace.	Type 5 (The Investigator), Type 9 (The Peacemaker)	Helps Type 5 detach from mental overactivity; centers Type 9.
Open up to Prayers	Use affirmations or prayers to connect with gratitude and higher meaning.	Type 2 (The Helper), Type 6 (The Loyalist)	Strengthens Type 2's spiritual compassion; provides reassurance for Type 6.
Open up to Your Dreams	Reflect on and interpret dreams to uncover unconscious patterns.	Type 4 (The Individualist), Type 5 (The Investigator)	Encourages self-discovery for Type 4; satisfies Type 5's curiosity.
Open up to Stargazing	Stargaze to connect with the universe and foster awe.	Type 7 (The Enthusiast), Type 4 (The Individualist)	Inspires wonder in Type 7; nurtures creativity for Type 4.
Open up to Entheogens	Explore safe & legal psychedelic experiences for transformative insight.	Type 1 (The Reformer), Type 5 (The Investigator)	Opens Type 1 to new perspectives; supports introspection for Type 5.

Open up to Mantras	Use chants or affirmations to foster focus and inner strength.	Type 2 (The Helper), Type 9 (The Peacemaker)	Centers Type 2; enhances calm and balance for Type 9.

Quick Ritual Reference Table

Ritual	1	2	3	4	5	6	7	8	9
1. Open up to Your Body								✓	✓
2. Open up to Your Breath	✓								✓
3. Open up to Your Sleep	✓		✓			✓			
4. Open up to Nature				✓					✓
5. Open up to Your Food		✓	✓						
6. Open up to Your Pets		✓				✓			
7. Open up to Yoga				✓	✓				✓
8. Open up to Dancing		✓					✓		
9. Open up to Running	✓							✓	
10. Open up to Evocative Music				✓					✓
11. Open up to an Instrument					✓		✓		
12. Open up to Laughter		✓					✓		
13. Open up to Uplifting Movies			✓			✓			
14. Open up to Gardening				✓					✓
15. Open up to Cooking		✓				✓			
16. Open up to Board Games			✓		✓				
17. Open up to Non-violent Video Games							✓		✓
18. Open up to Learning			✓		✓				
19. Open up to Traveling				✓			✓		
20. Open up to Art				✓	✓				
21. Open up to Meditation					✓				✓
22. Open up to Prayers		✓				✓			

23. Open up to Your Dreams				✓	✓				
24. Open up to Stargazing				✓			✓		
25. Open up to Entheogens	✓				✓				
26. Open up to Mantras		✓							✓

Recommended Mantra by Type

Type	Mantra	Why This Mantra Works
Type 1: The Reformer	*"I am enough just as I am."*	Addresses their inner critic and perfectionist tendencies, fostering self-acceptance.
Type 2: The Helper	*"My needs are just as important as others' needs."*	Encourages self-care and reduces over-dependence on serving others for validation.
Type 3: The Achiever	*"I am worthy beyond my accomplishments."*	Helps them detach self-worth from achievements and embrace authenticity.
Type 4: The Individualist	*"I am whole and complete as I am."*	Counters feelings of inadequacy and supports emotional grounding.
Type 5: The Investigator	*"I trust that I have all I need within me."*	Encourages emotional openness and reduces over-reliance on intellectual detachment.
Type 6: The Loyalist	*"I am safe and supported in this moment."*	Provides reassurance and helps them reduce anxiety and fear-driven behaviors.
Type 7: The Enthusiast	*"I find joy in stillness and presence."*	Balances their tendency to seek constant stimulation by cultivating mindfulness.
Type 8: The Challenger	*"Strength comes from allowing vulnerability."*	Encourages emotional openness and softens their defense mechanisms.

Type 9: The Peacemaker	*"My voice and needs matter."*	Supports assertiveness and encourages active participation in their own lives.

OPEN UP TO YOUR TYPE

A Journey to Wholeness and Transcendence

The journey of self-awareness through the Enneagram culminates in the profound act of opening up to your type. This is not merely about understanding your habitual patterns but transcending them to access the higher essence of who you are. At the heart of this transformation lies the *Holy Idea*—a spiritual or archetypal truth that each type longs to realize.

The concept of Holy Ideas originates from the spiritual tradition of the Enneagram, particularly the work of Oscar Ichazo, the founder of the modern Enneagram system. Ichazo, a Bolivian mystic and philosopher, introduced the Holy Ideas as part of his teachings on the Enneagram in the mid-20th century. He framed the Holy Ideas as metaphysical insights tied to the nine points of the Enneagram, representing the essential qualities of the divine or universal reality. Each Holy Idea is paired with a corresponding virtue—a behavioral expression of the Holy Idea—that reflects a type's potential for spiritual growth.

Some scholars believe that an early form of the Enneagram focused on seven passions as core ego fixations based on the Seven Deadly Sins (pride, envy, wrath, sloth, greed, gluttony, and lust). Ichazo reframed these sins (human flaws) as passions—deep emotional habits that shape personality. The list of passions in the Enneagram model closely parallels the Seven Sins but with an expanded psychological dimension:

Type	Passion (Ego Fixation)	Related to Seven Sins?
Type 1: Reformer	Anger & Resentment	Wrath
Type 2: Helper	Pride	Pride
Type 3: Achiever	Deceit (Self-Delusion)	(New Addition)
Type 4: Individualist	Envy	Envy
Type 5: Investigator	Avarice (Hoarding)	Greed
Type 6: Loyalist	Fear	(New Addition)
Type 7: Enthusiast	Gluttony	Gluttony
Type 8: Challenger	Lust (Excess)	Lust
Type 9: Peacemaker	Sloth (Avoidance)	Sloth

The transition from seven to nine types likely occurred as the system developed to include more nuanced psychological and spiritual insights. It also reflects a significant shift from a purely moralistic framework to a more psychological and spiritual one.

Ichazo's teachings were later expanded by students like Claudio Naranjo, who brought psychological depth to the Enneagram by linking the Holy Ideas to the types' mental fixations and defenses. This blend of spirituality and psychology has since become central to many Enneagram frameworks. The Holy Idea is a guiding light, inviting individuals to move beyond their ego-driven fears and desires into alignment with a greater sense of purpose and harmony.

The Holy Idea of Each Type

Opening your type is ultimately a path of transcendence, a movement from the narrow confines of personality toward the expansive possibilities of wholeness. It is an act of liberation, allowing you to embody the Holy Idea of your type and access the virtues and wisdom that lie beyond your habitual patterns. Engaging with your Holy Idea and embracing transcendence leads to the profound integration and freedom that define an *OPEN* life. It reflects universal truths about your life, connection, and existence that your type may distort or forget in its ego-driven state.

For example:

- **Type 1's Holy Perfection** reflects the divine truth that everything is inherently whole and perfect, countering their compulsion to fix flaws.
- **Type 6's Holy Faith** embodies the deep trust in the universe's support, helping them overcome anxiety and fear of instability.

The Holy Idea serves as a map to help you reconnect with your essence and break free from the limiting stories your personality has constructed.

Purpose of Your Holy Idea

Your Holy Idea is an abstract spiritual concept and practical tool for growth. It:

1. **Illuminates the Path of Transformation**: Showing you can align with their higher nature.
2. **Reveals the Ego's Illusions**: Helping you understand how your ego distorts reality.
3. **Offers Hope and Perspective**: Encouraging you to move beyond your fears and defenses into inner peace and wholeness.

Holy Idea for Each Type

This next table describes the Holy Idea for each type.

Type	Holy Idea	Description
Type 1: The Reformer	**Holy Perfection**	Recognizing that perfection already exists in the world and themselves, without needing to fix everything.
Type 2: The Helper	**Holy Will**	Understanding that love and care are abundant and not dependent on earning it by helping others.
Type 3: The Achiever	**Holy Hope**	Trusting their worth is intrinsic, not tied to external achievements or validation.
Type 4: The Individualist	**Holy Origin**	Realizing their connection to the universal whole, dissolving feelings of separation or lack.
Type 5: The Investigator	**Holy Omniscience**	Embracing the idea that true knowledge comes from unity and connection, not detachment.
Type 6: The Loyalist	**Holy Faith**	Finding security in trusting themselves and the universe rather than external systems or authorities.

Type	Holy Idea	Description
Type 7: The Enthusiast	**Holy Wisdom**	Recognizing that true joy and fulfillment arise from being fully present, not from seeking endless options.
Type 8: The Challenger	**Holy Truth**	Understanding that true strength comes from alignment with universal truth, not domination or control.
Type 9: The Peacemaker	**Holy Love**	Realizing that their presence and essence contribute to harmony and that they don't need to disappear to create peace.

Seeking A Sacred Space to OPEN Your Type

According to the teachings of *OPEN*, a sacred space can foster neuroplasticity and consciousness expansion to cultivate profound healing and self-discovery conditions. The concept of a sacred space, or *temenos*, is rooted in ancient traditions and enriched by Carl Jung's psychological insights. In this container, personal growth, change, and transcendence can occur. Jung emphasized that such physical or psychological spaces provide the safety and structure needed for deep inner work.

The Historic Significance of Temenos

In ancient Greek culture, temenos referred to a sacred precinct set apart for worship or communion with the divine. These spaces were often temples, groves, or sanctuaries consecrated to deities, where individuals could seek spiritual renewal or guidance. The boundaries of a temenos were marked clearly, symbolizing a threshold between the ordinary world and the sacred. Crossing into a temenos was not merely a physical act but a psychological and spiritual one, signifying readiness for transformation.

Carl Jung borrowed this ancient concept to describe the inner and outer spaces necessary for *individuation*—the process of integrating the conscious and unconscious aspects of the psyche. For Jung, a temenos was a symbolic space where the ego's defenses could be set aside, allowing suppressed thoughts, emotions, and archetypal energies to emerge and be processed. He saw temenos as a container for the often chaotic and vulnerable process of self-discovery, where one could confront their shadow and move toward wholeness.

Carl Jung on Psychic Safety

Jung emphasized that a temenos is essential for creating the psychic safety required to explore the unconscious. He likened temenos to the alchemical *vas*, or vessel, used in transformation. The vessel's boundaries were necessary to hold the volatile elements of the psyche during the change process. In therapy, Jung considered the therapeutic relationship a form of a *temenos*, where the dialogue between analyst and patient could unfold in a secure and sacred context.

Jung also linked *temenos* to the archetype of the sacred circle or mandala, representing order amidst chaos. He believed creating *temenos* could stabilize the psyche during tumultuous periods, allowing individuals to integrate fragmented aspects of themselves. According to Jung, such spaces are crucial for personal healing and spiritual transcendence, as they bridge the conscious self and the greater unconscious or collective psyche.

Creating Temenos*:* A Sacred Space for Transformation

The teachings of *OPEN* expand on Jung's ideas by emphasizing the practical creation of a temenos to foster neuroplasticity and higher consciousness. A sacred space, whether physical or symbolic, becomes a refuge where individuals can safely engage in rituals that rewire their brains, rewrite personal stories, and elevate their awareness. Your *temenos* thus can become a foundation for change and growth.

Characteristics of Your Temenos

1. **Boundaries**: The space must have clear physical, symbolic, or energetic boundaries to signify a departure from the ordinary and entry into the sacred. This separation enhances focus and intention.
2. **Safety**: your temenos must evoke a sense of security, enabling individuals to confront vulnerabilities and suppress emotions without fear.
3. **Intention**: Every ritual or action within your temenos should be guided by clear intentions aligned with healing, growth, or transcendence.
4. **Symbols and rituals**: Including meaningful symbols, artifacts, or rituals can anchor the space in personal and spiritual significance.

Rituals Within Your Temenos

Rituals performed within the sacred space amplify the transformative potential of your *temenos*.

Examples of *Temenos* in practice:

1. **The personal sanctuary**: A quiet corner in one's home, adorned with meaningful objects such as candles, crystals, or artwork, can serve as a personal temenos. This space becomes a refuge for daily meditation, journaling, or introspection.
2. **Natural settings**: Nature inherently provides a temenos. Forests, beaches, or gardens offer a sense of awe and connection, ideal for rituals like walking meditations or breathwork.
3. **Therapeutic contexts**: A therapist's office or a retreat center can act as a temenos where guided exploration of the psyche occurs within the safety of a professional relationship.
4. **Group settings**: Sacred circles, retreats, or ceremonial gatherings can create collective temeno*s* spaces, fostering communal healing and growth.

Temenos serves as a crucible for transformation. You can confront your fears, explore suppressed emotions, and rewrite disempowering narratives. This process is deeply aligned with the goals of *OPEN*, which emphasize the integration of body, mind, and spirit for holistic healing. In the sacred space, time seems to slow, allowing you to connect with your true self and the greater unconscious. Engaging in rituals that foster neuroplasticity heals old wounds and lays the foundation for sustained growth and resilience.

Temenos provides the ideal environment for this process. As Carl Jung noted, a sacred space is essential for creating the safety and focus needed to confront the shadow aspects of the psyche. Similarly, the Enneagram invites individuals to explore their type's strengths and vulnerabilities. Within the protective boundaries of a temenos, you can engage with your type's underlying fears and desires without judgment, allowing you to move toward a more integrated and open state.

The Significance of Opening Up To Your Type

In summary, here are the core benefits of opening up to your type.

1. **Moving Beyond the Ego**

Each Enneagram type is, in essence, a survival strategy shaped by ego defenses. For example:

- A Type 4 may cling to emotional intensity and uniqueness to avoid feelings of inadequacy.
- A Type 8 may assert control to mask vulnerability.
- A Type 6 may rely on loyalty to external systems to manage their fear of uncertainty.

Opening up your type means loosening these ego-driven patterns, allowing for a broader, more flexible sense of self. This process becomes possible in a *temenos* because you are held in a space where your defenses can be safely examined and softened.

2. **Integrating the Shadow**

Jung emphasized the importance of integrating the shadow—the repressed or unconscious aspects of the self. Each Enneagram type has a shadow side that reflects its unexamined fears and compulsions. For example, Type 1s might struggle with repressed anger, while Type 3s may bury their feelings of inadequacy under relentless achievement.

In the sacred space of a *temenos*, these shadow elements can surface and be acknowledged without fear or judgment. This integration allows you to transcend the limitations of their type, embracing their full humanity and potential.

3. Embodying Higher Virtues

The Enneagram teaches that each type has a higher essence or virtue that emerges when its patterns are transcended.

For instance:

- Type 4's melancholy transforms into equanimity that fosters inner peace, enhances decision-making, and supports personal growth by enabling individuals to respond to life's challenges with clarity and grace.
- Type 8's need for control evolves into true strength and vulnerability.
- Type 6's fearfulness becomes courage and faith.

Opening your type means accessing these higher virtues, which bring balance and freedom from the compulsions of personality. A *temenos* provides the setting for rituals and practices encouraging this shift, such as meditation, storytelling, or mindfulness.

4. **Aligning with Flow and Transcendence**

The teachings of *OPEN* emphasize achieving a state of flow—where the mind and body harmonize in an optimal state of engagement. By opening your type, you align with this flow, stepping out of rigid patterns into a more dynamic and adaptive way of being. This process is amplified in a temenos, where rituals and neuroplastic practices help rewire the brain and foster new patterns of openness and joy.

Towards a Neurospiritual Assessment of Stress, Anxiety, Depression, Addiction and Trauma

The Enneagram provides profound insights into each type's psychological tendencies and vulnerabilities. When viewed through this lens, the Enneagram can help illuminate how core fears, motivations, and habitual patterns might predispose individuals to specific psychological pathologies. This understanding fosters greater self-awareness and highlights potential areas for personal growth and healing.

The following tables explore the most likely psychological pathologies each Enneagram type may encounter and their propensity for experiencing SADAT conditions (Stress, Anxiety, Depression, Addiction, and Trauma). These insights combine research-supported findings with speculative connections to help individuals better understand their unique challenges and opportunities for growth.

While the Enneagram is not a diagnostic tool, its framework offers a valuable perspective for recognizing and addressing psychological patterns that might otherwise remain unconscious. Remember that not every individual of a particular type will face these challenges, and personal experiences may vary. Self-compassion, intentional action, and support from therapeutic or spiritual practices can significantly aid in addressing these tendencies. Explore the tables below to identify patterns relevant to your type and learn how they intersect with mental health concerns.

The insights provided in this section, including the tables on psychological pathologies and SADAT conditions, are intended for educational and self-reflective purposes only. The author of this guide is not a medical doctor, psychiatrist, or licensed therapist. These interpretations are based on research, personal experience, and the Enneagram framework and should not be used as a

substitute for professional medical advice, diagnosis, or treatment. If you or someone you know is experiencing significant mental health challenges, please consult a qualified healthcare professional or mental health specialist.

Always approach this material with self-compassion as a complementary resource to informed medical and psychological care.

<u>Link Between Pathologies and Type</u>

The following table outlines the most likely psychological pathologies that each Enneagram type may encounter, along with their propensity for experiencing SADAT conditions (Stress, Anxiety, Depression, Addiction, Trauma). Sources are referenced in the citation section; speculative insights are noted.

Type	Likely Psychological Pathologies	Likelihood of SADAT Conditions*	References/Notes
Type 1: The Reformer	Obsessive-Compulsive Personality Disorder (OCPD), Perfectionism, Self-Criticism	**Stress:** High **Anxiety:** Moderate **Depression:** Moderate	Their inner critic drives stress; OCPD link supported by Enneagram literature.
Type 2: The Helper	Codependency, Burnout, Emotional Overwhelm	**Stress:** High **Anxiety:** Moderate **Depression:** Moderate **Addiction:** Possible	Their focus on others can lead to emotional depletion; codependency noted in speculative sources.
Type 3: The Achiever	Workaholism, Narcissistic Traits, Performance-Driven Anxiety	**Stress:** High **Anxiety:** High **Depression:** Possible	Research supports burnout risks in overachievers.
Type 4: The Individualist	Mood Disorders (e.g., Bipolar Tendencies), Melancholy, Identity Issues	**Stress:** Moderate **Anxiety:** Moderate **Depression:** High **Addiction:** Possible	High depression risk due to introspection; speculative link to bipolar tendencies in some studies.
Type 5: The Investigator	Social Withdrawal, Schizoid Tendencies, Intellectualization	**Stress:** Moderate **Anxiety:** High **Depression:** Possible	Anxiety stems from overthinking; risk of withdrawal in personality studies.

Type	Likely Psychological Pathologies	Likelihood of SADAT Conditions *	References/Notes
Type 6: The Loyalist	Generalized Anxiety Disorder (GAD), Paranoia, Trust Issues	**Stress:** High **Anxiety:** High **Trauma:** Moderate	GAD link is well supported due to fear and uncertainty as core motivations.
Type 7: The Enthusiast	ADHD Traits, Impulsivity, Hedonistic Addiction	**Stress:** Moderate **Anxiety:** Possible **Addiction:** High	High addiction risk due to escapism; speculative ADHD association.
Type 8: The Challenger	Anger Management Issues, Control Issues, Oppositional Defiant Traits	**Stress:** High **Trauma:** Possible	Trauma arises from childhood vulnerability; speculative association with anger issues in psychological literature.
Type 9: The Peacemaker	Avoidant Personality Traits, Passive Depression, Emotional Numbing	**Stress:** Low **Anxiety:** Moderate **Depression:** High	Depression risk due to avoidance of conflict; emotional numbing is well-documented in avoidant behaviors.

***Notes on SADAT conditions:**

1. **Stress**: May be high across most types, driven by internal conflicts or external pressures.
2. **Anxiety**: May predominantly affect types driven by fear or uncertainty (Types 6 and 5).

3. **Depression**: May be higher in types prone to introspection or withdrawal (Types 4 and 9).
4. **Addiction**: May be more common in types seeking escape (Types 2, 4, 7).
5. **Trauma**: May be higher for types with unresolved childhood vulnerability (Types 8, 6).

The Enneagram has been explored in psychological research to understand its correlation with various psychological pathologies. Notable studies include:

1. **"The Enneagram: A Primer for Psychiatry Residents"** by Dr. David Daniels and Dr. Virginia Price (2015): This article discusses the application of the Enneagram in psychiatric practice, highlighting its potential in understanding personality disorders and therapeutic interventions.
2. **"Enneagram and Personality Disorders"** by Frédéric Schmitt (2014): This work examines the associations between Enneagram types and DSM-IV personality disorders, providing insights into how each type may correspond to specific psychological patterns.
3. **"The Enneagram as a Standard for the DSM"** by Elizabeth Wagele (2012): Published in *Psychology Today*, this article explores the potential of integrating the Enneagram framework with the Diagnostic and Statistical Manual of Mental Disorders (DSM) to enhance the understanding of personality disorders.

While these studies provide valuable insights, it's important to note that the Enneagram's integration into mainstream psychological practice is still evolving. Further empirical research must solidify its applications in diagnosing and treating psychological pathologies.

Correlating the Enneagram with the Big 5 Personality Model

The Big Five Personality Traits, also known as the Five-Factor Model, is a widely recognized framework in psychology that categorizes human personality into five broad dimensions:

1. **Openness to Experience**: Reflects imagination, creativity, and a willingness to consider new ideas.
2. **Conscientiousness**: Indicates organization, dependability, and discipline.
3. **Extraversion**: Pertains to sociability, assertiveness, and enthusiasm.
4. **Agreeableness**: Relates to trustworthiness, altruism, and cooperation.
5. **Neuroticism**: Involves emotional instability, anxiety, and moodiness.

The Big Five personality traits—Openness, Conscientiousness, Extraversion, Agreeableness, and Neuroticism—have been extensively validated through decades of research, establishing them as a robust framework for understanding human personality. The model gained prominence through the work of Paul T. Costa Jr. and Robert R. McCrae. Their development of the Revised NEO Personality Inventory (NEO PI-R) provided a comprehensive tool for assessing these traits. Their 1992 publication, "Revised NEO Personality Inventory (NEO PI-R) and NEO Five-Factor Inventory (NEO-FFI) Professional Manual," is a cornerstone in personality assessment.

Cross-cultural studies have supported the universality of the Big Five traits. McCrae and Allik's 2002 edited volume, "The Five-Factor Model of Personality Across Cultures," compiles research demonstrating the model's applicability across diverse cultures. Further, McCrae et al.'s 2005 study, "Personality Profiles of Cultures: Aggregate Personality Traits," analyzed data from 51 cultures, reinforcing the model's cross-cultural validity.

Psychometric Evaluations

Studies have examined the reliability and validity of Big Five assessments. For instance, a 2012 study titled "Confirmatory Factor Analysis on the Big 5 Personality Test Inventory" evaluated the internal consistency of the Big Five Inventory, reporting internal consistency coefficients of .80 for Extraversion, .75 for Agreeableness, .83 for Conscientiousness, .85 for Neuroticism, and .68 for Openness.

Neuroscientific Correlates

Research has explored the neural underpinnings of the Big Five traits. A 2007 study by DeYoung et al., "Testing Predictions From Personality Neuroscience: Brain Structure and the Big Five," found correlations between brain structure and personality traits, providing biological evidence supporting the model. Collectively, these studies underscore the extensive validation of the Big Five model, affirming its reliability and applicability across various contexts and cultures.

In the context of *OPEN*—a neurospiritual approach to mental health—the Big Five model offers valuable insights into how personality traits influence one's capacity for flow, resilience, and overall well-being. Understanding these traits can aid in identifying areas for personal growth and developing strategies to enhance mental health. The following table presents correlations between each Enneagram type and the Big Five personality traits based on available research and credible sources.

Table of correlation between Enneagram type and Big 5 traits

Enneagram Type	Openness	Conscientiousness	Extraversion	Agreeableness	Neuroticism
Type 1:	Moderate	High	Low to Moderate	Moderate	Moderate
Type 2:	Moderate	Moderate	High	High	Moderate
Type 3:	Moderate	High	High	Moderate	Low
Type 4:	High	Low to Moderate	Low	Low to Moderate	High
Type 5:	High	High	Low	Low	Moderate
Type 6:	Low to Moderate	High	Low to Moderate	High	High
Type 7:	High	Low	High	Moderate	Low
Type 8:	Low to Moderate	High	High	Low	Low
Type 9:	Moderate	Low to Moderate	Moderate	High	Moderate

Notes:

- **Openness**: Types 4, 5, and 7 often exhibit higher openness due to their creativity and curiosity.

- **Conscientiousness**: Types 1, 3, 5, and 6 tend to score higher, reflecting their organized and disciplined nature.
- **Extraversion**: Types 2, 3, 7, and 8 are generally more extroverted, while Types 4, 5, and 6 may lean towards introversion.
- **Agreeableness**: Types 2, 6, and 9 typically score higher, indicating their cooperative and empathetic tendencies.
- **Neuroticism**: Types 4 and 6 often exhibit higher levels associated with emotional sensitivity and anxiety.

Understanding these correlations can help individuals recognize inherent strengths and potential challenges, facilitating targeted strategies for personal development and mental health enhancement within the neurospiritual framework of *OPEN*.

While the Enneagram and the Big Five personality traits are two distinct frameworks used to understand human personality, ongoing research between these models continues to identify patterns and enhance the understanding of personality dynamics. Notable studies include:

1. **"The Enneagram and the Big Five Personality Traits"** by TraitLab (2021): This article examines the connections between the Enneagram types and the Big Five traits, providing insights into how each Enneagram type aligns with specific Big Five dimensions.
2. **"Correlation: Enneagram vs Big 5"** by @Prisme (2020): This study investigates the possible links between Enneagram types and the personality factors described in the Big Five model, highlighting predispositions of certain Enneagram types toward traits like extraversion or openness.
3. **"Enneagram vs. The Big Five"** by Scientific Enneagram (2018): This article summarizes at least nine studies examining the relationship between the Enneagram and the Big Five, providing a comprehensive overview of the research findings.
4. **"How do Enneagram types score on the Big Five personality traits?"** by Identity Abounds (2020): This analysis explores how different Enneagram types score on

the Big Five dimensions, offering a comparative perspective on personality assessments.

These studies contribute to a deeper understanding of how the Enneagram types correlate with the Big Five personality traits, offering valuable insights into both personal development and psychological research.

CONCLUSION

Sustaining an OPEN Life

OPEN UP is a transformative guide to understanding your unique challenges and opportunities for living a life of greater consciousness, plasticity, and flow. Each of the nine Enneagram types has specific traits, motivations, and fears that shape how they interact with the world. By recognizing these patterns, you gain the power to reprogram maladaptive behaviors, embrace openness, and move into a state of flow—a heightened, joyful experience where effort feels effortless, and life unfolds harmoniously.

Challenges and Opportunities: Harnessing the Enneagram for Growth

By now, you should be able to easily recognize that each Enneagram type carries intrinsic strengths and blind spots. These can fuel or hinder personal growth, depending on your level of self-awareness and engagement with neuroplastic practices. For example:

⇒ **GUT Types (8, 9, 1)**: These types grapple with control, anger, and autonomy issues. Type 8 must soften defenses to allow vulnerability, Type 9 needs to overcome inertia and disengagement to find assertiveness, and Type 1 must temper their perfectionism to embrace flexibility. Each challenge represents an opportunity for transformation, enabling these types to access their instincts with higher awareness and clarity.

⇒ **HEART Types (2, 3, 4)**: The HEART triad contends with fears of worthiness and emotional authenticity. Type 2 must balance their giving nature with self-care, Type 3 needs to align their ambition with a more profound purpose, and Type 4 must ground their introspection to avoid emotional overwhelm. Their journey is toward authentic connection, a key driver of flow.

⇒ **HEAD Types (5, 6, 7)**: Driven by the need for security and certainty, HEAD types must learn to embrace uncertainty and trust. Type 5 can cultivate openness by engaging emotionally, Type 6 can build trust by releasing anxiety-driven patterns, and Type 7 can deepen their life experience by practicing presence over escape. The HEAD types' opportunity lies in expanding their perspective and embracing curiosity.

By addressing these challenges and embracing opportunities, each type can achieve higher consciousness—a state where self-awareness leads to intentional, constructive action. Consciousness is about awareness of your thoughts and emotions and actively choosing the mindset and behaviors that align with your values and purpose.

The Role of Plasticity in Growth

Neuroplasticity, the brain's ability to rewire through experience, is central to personal transformation. By engaging with tailored rituals and practices, you create neural pathways that reinforce openness, flexibility, and flow. Each Enneagram type benefits from specific rituals designed to counteract its tendencies:

- ⇒ GUT types thrive with grounding practices like breathwork, physical activity, and nature immersion to reduce stress and balance their instinctual drives.
- ⇒ HEART types benefit from practices that uplift their emotions, like evocative music or laughter, but also mantras to foster self-awareness and authentic connections.
- ⇒ HEAD types can embrace practices like meditation, cognitive reappraisal, and experiential learning to build trust and adaptability.

The essence of plasticity is practice. Consistent engagement with rituals rewires your brain to sustain openness and transcend limiting patterns.

Rituals as Anchors for Openness and Flow

Rituals play a pivotal role in integrating the principles of *OPEN* into daily life. They anchor you in the present moment while fostering the mental flexibility and emotional resilience necessary for flow. Whether PRIMAL (connecting with the body), RATIONAL (engaging the mind), or SPIRITUAL (expanding the soul), rituals act as gateways to higher states of being.

Flow, the state of optimal engagement where you lose yourself in the task at hand, thrives on structure and rhythm. Embedding rituals into your life creates the conditions for flow to emerge. For example:

- ⇒ PRIMAL rituals like exercise or breathwork regulate the nervous system and prepare the body for openness.
- ⇒ RATIONAL rituals like problem-solving games or learning new skills engage the mind's focus and adaptability.
- ⇒ SPIRITUAL rituals like meditation, prayer, or stargazing cultivate inner peace and transcendence.

These practices synergize with your Enneagram type, addressing your unique challenges while fostering the openness required for flow.

The Journey Toward Flow: A Lifelong Commitment

Flow is the culmination of living an *OPEN life*. It is a state of profound connection where effort and outcome merge, producing joy, creativity, and a sense of purpose. For each Enneagram type, the path to flow requires a balance of consciousness and plasticity sustained by daily rituals. Achieving flow is not a one-time event but a dynamic, ongoing process of self-discovery and renewal.

Mihaly Csikszentmihalyi, the pioneer of flow research, says,

"The best moments in our lives are not the passive, receptive, relaxing times. The best moments usually occur if a person's body or mind is stretched to its limits in a voluntary effort to accomplish something difficult and worthwhile."

This resonates deeply with the Enneagram model, which encourages each type to embrace its challenges and opportunities to stretch toward its highest potential.

Living in flow means:

⇒ **Balancing Challenge and Skill**: Each type must identify where to push beyond their comfort zone without becoming overwhelmed.

⇒ **Cultivating Presence**: Flow requires being fully immersed in the moment, free from distractions or self-doubt.

⇒ **Sustaining Openness:** By engaging with rituals and neuroplastic practices, you maintain the flexibility and resilience to stay open to life's possibilities.

To live fully *OPEN*, you must embrace your Enneagram type with compassion and courage. Acknowledging your challenges and strengths is the first step toward raising your consciousness and enhancing your plasticity. Consistent rituals and practices lay the foundation for a life of flow—a state of harmony where your actions align with your essence.

Your Enneagram type is not a limitation but a roadmap to freedom. With the tools and insights *OPEN* provides, you can reprogram your mind, rewrite your story, and transform your life. Whether you are a bold Challenger, a reflective Individualist, a curious Enthusiast, or any other type, the journey toward openness and flow begins with one intentional step forward. Let this guide be your companion as you enter a future of greater joy, purpose, and fulfillment.

Flow is waiting for you. All you need to do is *OPEN UP!*

REFERENCES

Key Works by Dr. Christophe Morin on Neuroplasticity and Mental Health

1. Morin, C. (2025). *OPEN: A Neurospiritual Exploration of the Self-Healing Power of Your Brain.* Depth Insights.
2. Morin, C. (2020). *The Serenity Code: How Brain Plasticity Helps You Live Without Stress, Anxiety and Depression.* Depth Insights.

Foundational Enneagram Texts

3. Daniels, D., & Price, V. (2009). *The Essential Enneagram: The Definitive Personality Test and Self-Discovery Guide* (Revised ed.). HarperOne.
4. Palmer, H. (1991). *The Enneagram: Understanding Yourself and the Others In Your Life* (Revised ed.). HarperOne.
5. Riso, D. R., & Hudson, R. (1999). *The Wisdom of the Enneagram: The Complete Guide to Psychological and Spiritual Growth for the Nine Personality Types.* Bantam.
6. Riso, D. R., & Hudson, R. (1996). *Personality Types: Using the Enneagram for Self-Discovery* (Revised ed.). Houghton Mifflin Harcourt.
7. Rohr, R., & Ebert, A. (2001). *The Enneagram: A Christian Perspective.* Crossroad.
8. Chestnut, B. (2013). *The Complete Enneagram: 27 Paths to Greater Self-Knowledge.* She Writes Press.
9. Howe-Murphy, R. (2020). *Deep living with the Enneagram: Recovering your true nature (Revised and updated edition).* Deep Living Institute.
10. Howe-Murphy, R. (2007). *Deep coaching: Using the Enneagram as a catalyst for profound change.* Wisdom Way Press

Neuroscience and Psychology

11. Siegel, D. J. (2020). *The Developing Mind: How Relationships and the Brain Interact to Shape Who We Are.* Guilford Press.
12. Brown, B. (2010). *The Gifts of Imperfection.* Hazelden Publishing.
13. Sikora, M. (2022). *Enhancing Enneagram Therapy with Contemporary Research on the Conscious. SpringerLink.*
14. Edwards, A. C. (1991). *Clipping the wings of the Enneagram: A study of people's perceptions of a ninefold personality typology. Personality and Individual Differences, 12*(6), 613–618.
15. Furnham, A., & Crump, J. (2014). *The Eysenck Personality Profiler: An examination of the factor structure and scale reliabilities. Personality and Individual Differences, 66*, 81–85

Spiritual and Historical Perspectives

16. Gurdjieff, G. I. (1963). *Meetings with Remarkable Men.* Penguin Modern Classics.
17. Ichazo, O. (n.d.). Contributions documented in the Arica School materials.
18. Maitri, S. (2001). *The Spiritual Dimension of the Enneagram: Nine Faces of the Soul.* TarcherPerigee.
19. Naranjo, C. (1990). *Ennea-Type Structures: Self-Analysis for the Seeker.* Gateways Books.
20. Naranjo, C. (2004). *The Enneagram of Society: Healing the Soul to Heal the World.* Gateways Books & Tapes.

Comprehensive Guides

21. Heuertz, C. L. (2017). *The Sacred Enneagram: Finding Your Unique Path to Spiritual Growth.* Zondervan.
22. Cron, I. M., & Stabile, S. (2016). *The Road Back to You: An Enneagram Journey to Self-Discovery.* InterVarsity Press.

23. Stabile, S. (2018). *The Path Between Us: An Enneagram Journey to Healthy Relationships.* InterVarsity Press.

Behavioral and Practical Applications

24. Lapid-Bogda, G. (2004). *Bringing Out the Best in Yourself at Work: How to Use the Enneagram System for Success.* McGraw-Hill Education.
25. Lapid-Bogda, G. (2007). *What Type of Leader Are You? Using the Enneagram System to Identify and Grow Your Leadership Strengths and Achieve Maximum Success.* McGraw-Hill.
26. Wagner, J. (2010). *Nine Lenses on the World: The Enneagram Perspective.* Trafford Publishing.
27. Wagele, E. (1994). *The Enneagram Made Easy: Discover the 9 Types of People.* HarperOne.
28. Baron, R., & Wagele, E. (1995). *Are You My Type, Am I Yours? Relationships Made Easy Through the Enneagram.* HarperOne.

Enneagram Subtypes and Advanced Theory

29. Chestnut, B. (2017). *The 9 Types of Leadership: Mastering the Art of People in the 21st Century Workplace.* Post Hill Press.
30. Howell, J. (2021). *Enneagram Subtypes and Countertypes: A Brief Guide to 27 Distinct Versions of the Enneagram.* Enneagrammer Press.

Practical Applications in Life and Work

31. Goldsmith, M., & Fitzpatrick, R. (2004). *Awareness to Action: The Enneagram, Emotional Intelligence, and Change.* University of Scranton Press.

Enneagram and Psychiatry

32. Alexander, M., & Schnipke, B. (2020). The Enneagram: A primer for psychiatry residents. *The American Journal of Psychiatry Residents' Journal*, 15(3), 2–5.
33. *Citation:* Schmitt, F. (2014, September 4). Enneagram and personality disorders. *Institut M.A.R.I.E.* Retrieved from https://blog.enneagramme-marie.fr/en/blog/2014/09/04/enneagram-and-personality-disorders/
34. Wagele, E. (2012, December 5). The Enneagram as a standard for the DSM. *Psychology Today*. Retrieved from https://www.psychologytoday.com/us/blog/the-career-within-you/201212/the-enneagram-standard-the-dsm

Big Five Traits

35. Costa, P. T., Jr., & McCrae, R. R. (1992). Revised NEO Personality Inventory (NEO PI-R) and NEO Five-Factor Inventory (NEO-FFI) Professional Manual. Odessa, FL: Psychological Assessment Resources.
36. McCrae, R. R., & Allik, J. (Eds.). (2002). The Five-Factor Model of Personality Across Cultures. Springer Science & Business Media.
37. McCrae, R. R., Terracciano, A., & 78 Members of the Personality Profiles of Cultures Project. (2005). Personality profiles of cultures: Aggregate personality traits. *Journal of Personality and Social Psychology, 89*(3), 407–425. https://doi.org/10.1037/0022-3514.89.3.407
38. Hapsari, W., & Widhiarso, W. (2012). Confirmatory factor analysis on the Big 5 personality test inventory. *Journal of Educational and Developmental Psychology, 2*(1), 27-40. Retrieved from files.eric.ed.gov.
39. Soto, C. J., & John, O. P. (2024). Validation of an abbreviated Big Five personality inventory at large population scale. *Journal of Research in Personality, 97*, 101926. Retrieved from knowledge.uchicago.edu.

40. DeYoung, C. G., Hirsh, J. B., Shane, M. S., Papademetris, X., Rajeevan, N., & Gray, J. R. (2007). Testing predictions from personality neuroscience: Brain structure and the Big Five. *Psychological Science, 21*(6), 820–828. https://doi.org/10.1111/j.1467-9280.2007.01984.x

Studies Correlating Enneagram Types with Big Five Traits

41. Brown, A., & Bartram, D. (2009). *Relationships between the OPQ32i and the Enneagram.* SHL Group Limited. This study examines the connections between the Occupational Personality Questionnaire (OPQ32i) and Enneagram types, providing insights into how these frameworks relate. Retrieved from identityabounds.com
42. Newgent, R. A., Parr, P. E., Newman, I., & Higgins, K. K. (2004). *The Riso-Hudson Enneagram Type Indicator: Estimates of reliability and validity. Measurement and Evaluation in Counseling and Development, 36*(4), 226–237. This research assesses the reliability and validity of the Riso-Hudson Enneagram Type Indicator (RHETI) and explores its correlations with the Big Five personality traits.
43. Newgent, R. A., Parr, P. E., Newman, I., & Higgins, K. K. (2000). *The Enneagram: Trends in validation. Measurement and Evaluation in Counseling and Development, 33*(3), 166–175. This study reviews validation trends of the Enneagram and investigates its relationship with established personality models, including the Big Five.
44. TraitLab. (2021). *The Enneagram and the Big Five Personality Traits.* This article examines the connections between the Enneagram types and the Big Five traits, providing insights into how each Enneagram type aligns with specific Big Five dimensions. Retrieved from TraitLab.
45. Prisme. (2020). *Correlation: Enneagram vs Big 5.* This study investigates the possible links between Enneagram types and the personality factors described in the Big Five model, highlighting predispositions of certain Enneagram types toward traits like extraversion or openness.

46. Scientific Enneagram. (2018). *Enneagram vs. The Big Five.* This article summarizes at least nine studies that have examined the relationship between the Enneagram and the Big Five, providing a comprehensive overview of the research findings.
47. Identity Abounds. (2020). *How do Enneagram types score on the Big Five personality traits?* This analysis explores how different Enneagram types score on the Big Five dimensions, offering a comparative perspective on personality assessments.

ABOUT THE AUTHOR

Christophe Morin holds an MA and Ph.D. in Media Psychology from Fielding Graduate University and an MBA from Bowling Green State University. Dr. Morin is a distinguished lecturer at Johns Hopkins Whiting School of Engineering and an Adjunct Faculty at Fielding Graduate University, where he teaches several postgraduate courses on Prompt Engineering, Generative AI, the neuroscience of media effect, and neuromarketing.

Dr. Morin has over 30 years of experience in consumer, advertising, and public health research using cognitive psychology, media psychology, personality psychology, and neuroscience. He has authored three best-sellers on neuromarketing and scientific persuasion, currently available in multiple languages. He earned prestigious speaking, research, and publication awards during his remarkable career. During the COVID crisis, Dr. Morin pivoted his focus on the neurospiritual basis of mental health. In 2020, he published *The Serenity Code*, proposing a scientific formula for achieving more serenity through self-love habits.

Since then, Dr. Morin has delved deeper into neuroplasticity's materialistic and non-materialistic foundations, exploring its powerful connection to transformation and self-healing. In November 2023, he embarked on a month-long shamanic initiation retreat in Peru with the Shipibo tribe. This experience inspired him to begin writing *OPEN*, his most personal work. The book combines cutting-edge research on the neuroscience of self-healing with ancient traditions and the responsible, safe use of entheogens to help alleviate or eliminate these conditions.

INDEX

Made in United States
North Haven, CT
04 November 2026